Richard Anderson

Lightning Conductors

Their History, Nature, and Mode of Application

Richard Anderson

Lightning Conductors
Their History, Nature, and Mode of Application

ISBN/EAN: 9783337250676

Printed in Europe, USA, Canada, Australia, Japan

Cover: Foto ©berggeist007 / pixelio.de

More available books at **www.hansebooks.com**

LIGHTNING CONDUCTORS

THEIR

HISTORY, NATURE, AND MODE OF APPLICATION

BY

RICHARD ANDERSON, F.C.S. F.G.S.

MEMBER OF THE SOCIETY OF TELEGRAPH ENGINEERS
ASSOC. INST. C. E.

WITH NUMEROUS ILLUSTRATIONS

LONDON
E. & F. N. SPON, 46 CHARING CROSS
NEW YORK
446 BROOME STREET
1880

PREFACE.

THE WANT in England of a good practical work on Lightning Conductors, accessible to both the professional and non-professional reader, has long been a subject of remark. That there are English works bearing more or less on Lightning Protection will be seen at once on reference to the Bibliography contained in the Appendix, pp. 231–248. But it will be found these books are either obsolete and out of print, or are written in a purely popular style that conveys little or no 'usable' information whereby may be obtained a trustworthy account of the growth and application of the LIGHTNING CONDUCTOR.

It is with a view of meeting this need that the present work has been written. It contains not only a history of the various methods that have been used to this end, but also a thoroughly practical exposition of the systems employed by the best authorities in various countries.

To Architects, Clergymen, Municipal Officials, and all those in charge of large and lofty buildings, it would be impossible to over-estimate the importance of this subject. Year by year an enormous amount of property is destroyed merely because the simplest precautions have not been taken to guard churches and other large buildings from the effects of thunder storms.

The Author of this work can at all events claim a large practical acquaintance with its subject. He feels convinced that those concerned in the preservation of buildings, whether they be houses, churches, or public offices, need only to learn the simple methods that can be used to render the action of lightning innocuous, in order to adopt them.

R. A.

NEW MALDEN, SURREY:
October 1879.

CONTENTS.

CHAPTER		PAGE
	LIST OF BOOKS REFERRED TO, OR CONSULTED, RELATING TO LIGHTNING CONDUCTORS	xi
I.	ELECTRICITY AND LIGHTNING .	1
II.	DISCOVERY OF THE LIGHTNING CONDUCTOR	17
III.	EARLY EXPERIMENTS WITH LIGHTNING CONDUCTORS . .	25
IV.	GRADUAL SPREAD OF LIGHTNING CONDUCTORS IN EUROPE . .	34
V.	METALS AS CONDUCTORS OF ELECTRICITY	49
VI.	CHARACTER OF LIGHTNING AND OF THUNDERSTORMS . .	62
VII.	INQUIRIES INTO LIGHTNING PROTECTION	73
VIII.	SIR WILLIAM SNOW HARRIS	85
IX.	THE BEST MATERIAL FOR CONDUCTORS .	100
X.	HOTEL DE VILLE, BRUSSELS, AND WESTMINSTER PALACE	111
XI.	WEATHERCOCKS . . .	121
XII.	LIGHTNING PROTECTION IN FRANCE AND AMERICA	125
XIII.	NEWALL'S SYSTEM OF PROTECTING BUILDINGS . .	140
XIV.	ACCIDENTS AND FATALITIES FROM LIGHTNING	169
XV.	THE EARTH CONNECTION	198
XVI.	INSPECTION OF LIGHTNING CONDUCTORS	218
	APPENDIX . . .	231
	INDEX	249

LIST OF BOOKS

REFERRED TO, OR CONSULTED, RELATING TO THE HISTORY, NATURE, AND MODE OF APPLICATION OF

LIGHTNING CONDUCTORS.

ACHARD (Fr. K.) Kurze Anleitung ländliche Gebäude vor Gewitterschäden sicher zu stellen. 8vo. Berlin, 1798.

ARAGO (François). Meteorological Essays. Translated by Colonel Sabine; with an Introduction by Baron von Humboldt. 8vo. London, 1855.

BARBERET (J.) Dissertation sur le Rapport qui existe entre les Phénomènes de Tonnerre et ceux de l'Electricité. 2 vols. 4to. Bordeaux, 1750.

BEAUFORT (Dr. Antonin de). Notice sur les Paratonnerres. 8vo. Châteauroux, 1875.

BECCARIA (C. B.) Lettere dell' Elettricismo. 4to. Bologna, 1758.

BECCARIA (Giambatista). A Treatise upon Artificial Electricity. Translated from the Italian. 8vo. London, 1776.

BECQUEREL (Antoine C.) Traité de l'Electricité et du Magnétisme. 7 vols. 8vo. Paris, 1834–40.

BERGMAN (T.) Tal om möjeligheten at förexomma askans skadeliga werkningar. 4to. Stockholm, 1764.

BIGOT (P.) Anweisung zur Anlegung, Construction und Veranschlagung der Blitzableiter. 8vo. Glogau, 1834.

BODDE (M.) Grundzüge zur Théorie der Blitzableiter. 8vo. Münster, 1804.

BOECKMANN (N.) Ueber die Blitzableiter. 8vo. Karlsruhe, 1791.

BREITINGER (D.) Instruction über Blitzableiter. 4to. Zürich, 1830.

BUCHNER (Dr. Otto). Die Konstruction und Anlegung der Blitzableiter, mit einem Atlas. 2nd edition, 8vo. Weimar, 1876.

CALLAUD (A.) Traité des Paratonnerres. 8vo. Paris, 1874.

CAVALLO (Tiberius). A Complete Treatise on Electricity. 2 vols. 8vo. London, 1786.

LIST OF BOOKS CONSULTED.

DALIBARD (M.) Histoire abrégée de l'Electricité. 2 vols. 8vo. Paris, 1766.

DAVY (Sir Humphrey). Elements of Chemical Philosophy. 8vo. London, 1810.

DEMPP (K. W.) Vollständiger Unterricht in der Technik der Blitzableitersetzung. 8vo. München, 1842.

EBERHARD (Dr.) Vorschläge zur bequemeren und zicherern Anlegung der Pulvermagazine. 8vo. Halle, 1771.

EISENLOHR (Dr. W.) Anleitung zur Ausführung und Visitation der Blitzableiter. 8vo. Karlsruhe, 1848.

EITELWEIN (J. A.) Kurze Anleitung auf welche Art Blitzableiter an den Gebäuden anzulegen sind. 8vo. Berlin, 1802.

FAIT (E. M.) Observations concerning Thunder and Electricity. 8vo. Edinburgh, 1794.

FERGUSON (James). An Introduction to Electricity. 3rd edition, 8vo. London, 1778.

FIGUIER (Louis). Les Merveilles de la Science. 4to. Paris, 1867.

FONVIELLE (Wilfrid de). Eclairs et Tonnerres. 8vo. Paris, 1869.

FONVIELLE (Wilfrid de). De l'Utilité des Paratonnerres. 8vo. Paris, 1874.

FRANKLIN (Benjamin). Experiments and Observations in Electricity, made at Philadelphia, in America. 8vo. London, 1751.

FRANKLIN (Benjamin). Complete Works in Philosophy, Politics, and Morals. 3 vols. 8vo. London, 1806.

FRANKLIN (William Temple). Memoirs of the Life and Writings of Benjamin Franklin. 4to. London, 1818.

GAY-LUSSAC (F.) et POUILLET (Claude). Introduction sur les Paratonnerres, adoptée par l'Académie des Sciences. 8vo. Paris, 1874.

GRENET (E.) Construction de Paratonnerres. 8vo. Paris, 1873.

GROSS (J. F.) Grundsätze der Blitzableitungskunst. 8vo. Leipzig, 1796.

GUERICKE (Otto von). Experimenta nova Magdeburgica. Folio. Amsterdam, 1672.

GÜTLE (J. K.) Neue Erfahrungen über die beste Art Blitzableiter anzulegen. 8vo. Nürnberg, 1812.

HARRIS (William Snow) On the Nature of Thunderstorms, and the Means of Protecting Buildings and Shipping against Lightning. 8vo. London, 1843.

HARRIS (Sir William Snow). A Treatise on Frictional Electricity. Edited by Charles Tomlinson. 8vo. London, 1867.

HELFENZRIEDER (J.) Verbesserung der Blitzableiter. 8vo. Eichstadt, 1783.

HEMMER (J. J.) Kurzer Begriff und Nutzen der Blitzableiter. 8vo. Mannheim, 1783.

HEMMER (J. J.) Anleitung Wasserableiter an allen Gattungen vor Gebäuden auf die sicherste Art anzulegen. 8vo. Frankfurt, 1786.

HENLEY (William). Experiments concerning the Different Efficacy of Pointed and Blunt Rocks in securing Buildings against the Stroke of Lightning. 8vo. London, 1774.

HOLTZ (Dr. W.) Ueber die Theorie, die Anlage und die Prüfung der Blitzableiter. 8vo. Greifswald, 1878.

IMHOF (M. von). Theoretisch-practische Anweisung zur Anlegung zweckmässiger Blitzableiter. 8vo. München, 1816.

INGENHOUSZ (Dr. Johan). New Experiments and Observations concerning Various Subjects. 8vo. London, 1779.

KLEIN (Hermann J.) Das Gewitter und die dasselbe begleitenden Erscheinungen. 8vo. Graz, 1871.

KUHN (Carl). Handbuch der angewandten Elektricitätslehre. Part 1. Ueber Blitzableiter. 8vo. Leipzig, 1866.

LANDRIANI (M.) Dell' Utilità di Conduttori elettrici. 4to. Milano, 1785.

LAPOSTOLLE (M.) Traité des Parafoudres et des Paragrêles. 8vo. Amiens, 1820.

LENZ (Heinrich F. E.) Handbuch der Physik. 2 vols., 8vo. Petersburg, 1864.

LICHTENBERG (G. Ch.) Neueste Geschichte der Blitzableiter. 8vo. Leipzig, 1803.

LUTZ (F.) Unterricht vom Blitze und Wetterableitern. 8vo. Nürnberg, 1783.

MAFFEI (F. S.) Delle Formazione dei Fulmini. 4to. Verona, 1747.

MAHON (Lord). Principles of Electricity. 4to. Elmsly, 1780.

MARUM (M. van). Verhandeling over hat Electrizeeren. 8vo. Groningen, 1776.

MEISENS (M.) Notes sur les Paratonnerres, in 'Bulletins de l'Académie Royale de Belgique.' 8vo. Bruxelles, 1874-78.

MEISENS (M.) Des Paratonnerres. 4to. Bruxelles, 1877.

MEURER (Heinrich). Abhandlung von dem Blitze und den Verwahrungsmitteln gegen denselben. 4to. Trier, 1791.

MURRAY (N.) Treatise on Atmospheric Electricity, including Observations on Lightning Rods. 8vo. London, 1828.

NEWALL (R. S.) Lightning Conductors: their use as protectors of buildings, and how to apply them. 8vo. London, 1876.

NOAD (Henry M.) Lectures on Electricity. 8vo. London, 1849.

NOAD (Henry M.) A Manual of Electricity. 8vo. London, 1855.

LIST OF BOOKS CONSULTED.

NOLLET (Abbé J. A.) Leçons de Physique expérimentale. 6 vols. 12mo. Paris, 1743.

NOLLET (Abbé J. A.) Recherches sur les Causes particulières des Phénomènes électriques. 8vo. Paris, 1749.

OHM (Georg Simon). Bestimmung des Gesetzes nach welchem die Metalle die Contact-Electricität leiten. 8vo. Nürnberg, 1826.

PARTON (James). Life and Times of Benjamin Franklin. 2 vols. 8vo. New York, 1864.

PHIN (John). Plain Directions for the Construction of Lightning Rods. 8vo. New York, 1873.

PLIENINGER (Dr. P.) Ueber die Blitzableiter. 8vo. Stuttgart, 1835.

PONCELET (Abbé M.) La Nature dans la Formation du Tonnerre. 8vo. Paris, 1766.

POUILLET (Claude S. M.) Eléments de Physique expérimentale et de Météorologie. 7th edition, 2 vols., 8vo. Paris, 1856.

PRAIBSCH (Christian). Ueber Blitzableiter, deren Nutzbarkeit und Anlegung. 8vo. Zittau und Leipzig, 1830.

PREECE (W. H.) On Lightning and Lightning Conductors, in 'Journal of the Society of Telegraph Engineers.' 8vo. London, 1873.

PRIESTLEY (Dr. Joseph). The History and Present State of Electricity. 2 vols. 8vo. London, 1775.

REIMARUS (J. A. H.) Vom Blitze. 8vo. Hamburg, 1778.

REIMARUS (J. A. H.) Ausführliche Vorschriften zur Blitz-Ableitung an allerlei Gebäuden. 8vo. Hamburg, 1794.

ROBERTS (M.) On Lightning Conductors, particularly as applied to Vessels. 2 vols. 8vo. London, 1837.

ROWELL (G. A.) An Essay on the Cause of Rain and its Allied Phenomena. 8vo. Oxford, 1859.

SAUSSURE (H. B. de). Manifeste, ou exposition abrégée, de l'Utilité des Conducteurs électriques. 8vo. Genève, 1771.

SIGAUD DE LA FOND (M.) Précis historique et expérimental des Phénomènes électriques. 2nd edition, 8vo. Paris, 1785.

SINGER (George John). Elements of Electricity. 8vo. London, 1814.

SPANG (Henry W.) A Practical Treatise on Lightning Protection. 8vo. Philadelphia, 1877.

SPARKS (Jared). The Works of Benjamin Franklin; with Notes and a Life of the Author. 10 vols. 8vo. Boston, 1840.

SPRAGUE (John F.) Electricity: its Theory, Sources, and Applications. 8vo. London, 1875.

STRICKER (Dr. Wilhelm). Der Blitz und seine Wirkungen. 8vo. Berlin, 1872.

STURGEON (William). Lectures on Electricity. 8vo. London, 1842.

TAVERNIER (A. de). Blitzableiter, genannt Anti-Jupiter. 8vo. Leipzig, 1833.

TINAN (Barbier de). Mémoires sur les Conducteurs pour préserver les Edifices de la Foudre. 8vo. Strasbourg, 1779.

TOALDO (Giuseppe). Della Maniera di defendere gli Edifizii dal Fulmine. 8vo. Firenze, 1770.

TOALDO (Giuseppe). Dei Conduttori per preservare gli Edifizii da Fulmine. 4to. Venezia, 1778.

TOMLINSON (Charles). The Thunderstorm. 8vo. London, 1859.

TYNDALL (John). Notes on Electrical Phenomena. New edition, 8vo. London, 1876.

VERATTI (J.) Dissertatione de Electricitati coelèsti. 8vo. Bologna, 1755.

WEBER (F. A.) Abhandlung von Gewitter und Gewitterableiter. 8vo. Zürich, 1792.

WHARTON (W. L.) The Effect of a Lightning Stroke. 8vo. London, 1841.

WILSON (Robert). Boiler and Factory Chimneys; with a chapter on Lightning Conductors. 8vo. London, 1877.

WINCKLER (Prof. J. H.) Gedanken von den Eigenschaften, Wirkungen und Ursachen der Elektricität. 8vo. Leipzig, 1744.

YELIN (J. C. von). Ueber die Blitzableiter aus Messingdrahtstricken. 8vo. München, 1824.

LIGHTNING CONDUCTORS:

THEIR

HISTORY, NATURE, AND MODE OF APPLICATION.

CHAPTER I.

ELECTRICITY AND LIGHTNING.

'First let me talk with this philosopher: What is the cause of thunder?' asks Shakspeare in 'King Lear,' but without giving a reply. The 'philosopher' of Shakspeare's days had no answer to make; nor had any others long after. From the dawn of history till within comparatively modern times, thunder and lightning were mysteries to the human mind; nor did there exist so much as a surmise that there might be any connection between them and the equally mysterious agent called electricity. The latter force indeed revealed itself early to attentive observers, though in forms very different from those known at the present time. The Greeks found out that amber, or 'electron,' attracted certain other bodies under friction, and named the force after it; and the Romans were aware that the shocks discharged by the torpedo fish were of electrical nature, and they used them for the cure of rheumatic complaints in the reign of the Emperor Tiberius. Both Greeks and Romans also observed the sparks emitted, under certain circumstances, from clothing and from the fur of animals. But this represented the total sum of knowledge about electricity for ages and ages.

It was not until the year 1600 that Dr. William Gilbert, physician to Queen Elizabeth, made a great step forward by showing in his celebrated work, 'De magnete, magneticisque corporibus, et de magno magnete tellure, physiologia nova,' that the two classes of phenomena, the magnetic and the electric, are emanations of a single fundamental force pervading all nature. Dr. Gilbert further discovered that many other substances besides amber possess the electric power, and that this power is easily excited when the air is dry and cool, and with difficulty when it is moist and warm. These discoveries caused great commotion in the European learned world, yet produced no further result for another half a century. In 1650, Otto von Guericke, burgomaster of Magdeburg, the inventor of the air-pump, who had studied with deep interest Dr. Gilbert's book, succeeded in constructing a little electrical machine, composed mainly of a ball of sulphur mounted on a revolving axis. By the aid of this instrument, very rude in construction, he produced powerful sparks and flashes of electric light, and it helped him likewise to discover, first, that bodies excited by friction communicate their electricity to other bodies by mere contact, and, secondly, that there resides in electrified substances the power of repulsion as well as that of attraction.

Those who followed in the wake of the ingenious burgomaster of Magdeburg for the next ninety or hundred years, till towards the middle of the eighteenth century, did very little towards adding to the already acquired knowledge of electricity. Sir Isaac Newton constructed an electrical machine of glass, very superior to that of Otto von Guericke, with which he made some amusing experiments; but, strangely enough, drew no conclusions from them, treating the mighty force under his eyes as only a plaything. This was all the more singular as a contemporary of the great philosopher, Francis Hauksbee, like him a Fellow of the Royal Society, called attention, in a volume entitled 'Physico-mechanical Experiments,' published in 1709, to the great similarity between the electric flash and lightning, hinting that the two might possibly be offspring of the same mysterious force. Dr. Wall, in 1708, said that the light and crackling of rubbed amber seemed in some degree to represent thunder and lightning. Another member of the Royal Society, Stephen Gray—the first man in England who made the study of electricity the devotion of his life, but of whose career very little is known beyond the fact that he was very poor, and a pensioner of the Charterhouse—added numberless experiments to those previously made, and was bold enough to declare, in 1720, six years before Sir Isaac Newton's death, that '.electricity seems to be of the same nature with thunder and lightning—if we may compare great things with small.' For this audacity in 'comparing things' he was sharply taken to task by all the scientific men of the age, and, as deserved, set down as a man out of his senses.

Nothing more was done for the next twenty-five years to enlarge the knowledge of the phenomena of electricity. It stood, in fact, on a footing not very far advanced from what it had been two thousand years before. The achievements mainly consisted in a great number of entertaining experiments performed for the delectation of great and little children. Various machines had been made for exciting electricity, but they served only, or at least chiefly, for amusement, allowing ladies to fire off a cannon by a touch of their delicate hand, and bringing ladies and gentlemen together to behold the wonderful spectacle of an infant's hair being made to stand on end, the little creature having been placed upon cakes of resin, and fastened to the ceiling by silken cords. The whole was little more than a repetition, on a greater scale and with improved means, of the ancient Greek experiment of rubbing a piece of amber on the sleeve of a philosopher's coat.

The first great step towards a practical insight into the nature and phenomena of electricity, hitherto a mere plaything, was made in the year 1745 in the ancient Dutch city and university of Leyden. Two professors of the high school, John Nicholas Allamand, a member of the Royal Society of London, and Peter Van Musschenbroek, author of a treatise entitled 'Introductio ad philosophiam naturalem,' had been trying, like many other scientific men of the time, electrical experiments, when the thought occurred to them that the real reason why all the work of the same kind had as yet produced such slight results was that the electrical force was absolutely unstable. It slipped, so to speak, through their hands, before they could look at it; it vanished 'like a dream, leaving no substance behind.' One body, they knew, had the power of electrifying another, but only to let the mysterious force pass on, like a current of water running down a cataract. Could they but 'bottle up' electricity, what a grand gain would this be to science! So thought the two professors of Leyden university; and thought justly. They went on experimenting, with this end in view, till at last so-called 'accident,' the mother of millions of human inventions and discoveries, threw a brilliant light on the dark road along which they were groping their way.

One day Professor Allamand and Van Musschenbroek, together with a pupil named Cuneus—a sort of Wagner, it would seem, sitting at the feet of Dr. Faust—were trying the effects of electricity on a small iron cannon, suspended by silk threads, and connected by a wire with a glass bottle half full of water, when whey were startled by an extraordinary incident. Curious, like all students of occult sciences, young Cuneus took it into his head to see what would happen if he held the prime conductor of the electrical machine in one hand and the electrified bottle of water in the other. Something wonderful happened, indeed, causing profound amazement and terror to the three persons witnessing it, most of all to the immediate experimenter, who sank down on the floor, half dead with fright. Master Cuneus had received an electric shock. It was the first electric shock ever administered by artificial means to any human being.

Such was the origin of the long-famous 'Leyden jar,' or, as it was originally called, 'Leyden phial.' The whole of the scientific world of Europe was as much startled by the discovery that electricity could be imprisoned, like Ariel in an oak-tree, as the two Leyden professors and their pupil had been, and a perfect fury set in for more experiments. A professor of the University of Leipzig, in Germany, Dr. Winckler, started the excitement by submitting his body to frequent powerful shocks, opening up, besides, a scientific discussion in which he came forth as the champion of the proposition that the discovery of the 'Leyden phial' was due, not to the professors in the Dutch university, but to a German ecclesiastic, Ewald George von Kleist, who made the experiments of Messrs. Allamand and Van Musschenbroek a year before them. His own sensations in submitting to the force of electric shocks, Professor Winckler described, doubtless with some exaggeration, as being convulsed from head to toe, and the prey of violent agitations, which threw his arms about, and made the blood rush from his nose. Dr. Winckler did not venture upon many experiments; but his spouse, undismayed by the arm-shaking and nose-bleeding of her lord, and having the combined curiosity of a woman and a professor's wife, continued upon her own person the electric shocks. However, she did not take many, nor did science gain by the sacrifice. When a few graspings of the 'Leyden phial' had deprived her of the power to walk, and, what was worse, to speak, she followed the example of her bleeding husband, and took 'cooling medicines.' All these wonderful facts were made widely known at the time, and created the most profound interest. Professor Musschenbroek, of Leyden, added not a little to the prevailing excitement by writing to his friend René Antoine de Réaumur, inventor of the thermometer named after him, a long letter, given at once to the public, in which he dwelt upon the terrible effects of the mysterious agency which he had helped to call into being, and wound up by declaring that he had become terrified by his own foster-child, and that he would not submit to another electric shock 'for the whole kingdom of France.'

Experiments in electricity now became the prevailing mania. Louis XV. of France set the fashion among crowned heads of having his soldiers electrified, to see what benefit he, or they, would derive from it. On the instigation of Abbé Nollet, considered a man of high scientific attainments, and who made several important discoveries in electricity, the King submitted, in his own presence, 180 of the tallest men of his life-guards, fastened hand to hand by iron wires, to repeated charges from a connected group of Leyden jars. The big fellows were not visibly influenced by the electric shocks, experiencing not so much as the historical nose-bleeding of Professor Winckler of Leipzig, still less the dumbness of his worthy spouse. On the contrary, the wire-bound royal guards, conscious of but very slight sensations from the electric shocks, and feeling somewhat indignant at this, and of being made scientific tools without at least getting a strong bump on the head, spoke out strongly, declaring the whole matter to be an imposture.

Having failed to electrify his soldiers, Louis XV. tried his monks. It struck his Most Christian Majesty that perhaps the human creatures who had the honour of fighting for him were endowed by nature with rather tough hides, and that the case might be different in regard to the softer beings upon whom devolved the task of praying for him. Accordingly, the King issued orders that all the monks of the grand convent of the Carthusians at Paris, over 700 in number, should be electrified by the same connected group of Leyden jars which had been tried upon the company of life-guards. The result was entirely different, and most gratifying to the King. The shock had no sooner been given when the whole file of monks gave an instantaneous jump, uttering a howl at the same time. There were some eye-witnesses of the affair who asserted that the Carthusians jumped and howled even before the shock had been given, on seeing some one approach the Leyden jar;

but this was officially denied. King Louis XV. was so delighted with this result of his scientific investigations, that he proposed to submit all the monks of all the monasteries of France successively to the process of being electrified, so that it might be accurately ascertained upon what religious orders and communities it took the greatest effect. His Majesty likewise was pleased to suggest, that, after all the monks had been electrified, the nuns might be tried in their turn. But the proposal was vetoed at Rome. There came definite orders from the Supreme Pontiff forbidding the contact of any more persons in the service of the holy Catholic Church with the sinful electric wire; and the Carthusians of Paris remained the last monks, as they had been the first, brought to jump and howl at the touch of a Leyden jar.

From France and the continent of Europe the mania for electrical experiments spread into England. But here it was taken up in a thoroughly practical spirit, worthy of the genius of the nation. Instead of aiming merely at the production of wonderful phenomena, made to create astonishment, a number of scientific gentlemen formed themselves into a body for the express purpose of seeking to ascertain the nature, effects, and conditions of the mysterious agent which had obtained the name of electricity. At the head of this body of inquirers was Dr. William Watson, a member of the Royal Society, indefatigable in the pursuit of science, and with him worked Martin Folkes, then president of the Society, Lord Charles Cavendish, Dr. Bevis, and other distinguished men. They set themselves, first of all, to ascertain in what manner electricity was communicated through the solid earth, as well as through fluid bodies; and, secondly, to enter upon experiments showing the amount of speed at which the force travelled. With the first object before them, they made some curious trials in the month of July 1747, which attracted all London. They hung a wire over the Thames, close to Westminster Bridge, attaching the one end to a Leyden jar, and giving the other to a man who held it in the left hand, while he grasped with the right an iron rod, standing in the river. Facing the latter, on the opposite side of the Thames, not far from the operators with the 'jar,' was stationed another person, also grasping an iron staff planted in the river. After the charge had been given, it was found that the electricity, after travelling by the wire over the river, had come back by the water, the person holding the iron staff on the starting side not only experiencing a shock himself, but several individuals touching him. Not content with this experiment, showing the transmission of electricity, Dr. Watson and his friends made another, on a larger scale, a week afterwards, on the New River, at Stoke Newington, London. They spanned, by chains and wires, a circuit embracing 800 feet by land and 2,000 by water, with the result of finding that the water transmitted the electric force by itself, if merely an iron staff was placed in it. But they also discovered at the same time that moist land would carry the force, equally with water. To ascertain the latter fact more distinctly, the investigators made a third experiment at Highbury Barn, Islington, setting up some miles of wire, separated partly by land and partly by water. The conduit of the electric force throughout the whole distance was found to be uninterrupted, which led Dr. Watson to proclaim his conviction that the agent was far more abundant throughout nature than had been formerly believed.

In order to ascertain the speed at which the electric force traversed space, Dr. Watson and his friends next entered upon a series of experiments at Shooter's Hill, near London. They sent an electric discharge a distance of four miles, observers being stationed at each end, and a gun fired at the touch of the Leyden jar, when it was shown conclusively that the movement of the electric force was instantaneous. This was an important step in advance, in overthrowing all formerly established conclusions as to the agency being produced by a succession of waves, like sound, and as such, moving slowly through space.

The field for electrical experiments was now becoming

gradually more extensive, and a few more practical tests of Mr. Watson and his coadjutors led the way to the greatest knowledge of the all-pervading force that had yet been achieved, in the clear apprehension that lightning was but a manifestation of electricity. The new experiments were chiefly made with the so-called electrical tube, a glass rod, from two and a half to three feet in length and about an inch in diameter. It had been known for some time that the tube, when gently warmed, so as to be perfectly dry, and rubbed with a silk handkerchief, exhibited strong symptoms of electricity, to the extent of throwing off luminous sparks, which obtained the name of 'electric fire.' Dr. Watson found, to his surprise, that this electric fire was not general and always obtainable, but conditional upon circumstances. Having rubbed a glass tube while he was insulated by standing upon a cake of wax, he found that no electricity could be drawn from him by another person who touched any part of his body, but that the same person could obtain sparks from the tube by putting his hand near it. Dr. Watson likewise observed, in the same train of experiments, that if an electrical machine, together with the person turning the handle, were suspended by silk, electric fire was not apparent until he touched the floor with one foot, when the fire appeared upon the conductor. Having made a great number of trials of a like nature, Dr. Watson made known the important conclusion derived from them, namely, that glass tubes and all similar 'electrifiers' did not contain within themselves the subtle agent known as electricity, but formed only its temporary place of rest, as a sponge would that of water. Dr. Watson was near proclaiming the fact that electricity resides everywhere throughout the universe; but for a moment he only touched the fringe of it. The discovery of this grand truth was left to later investigators.

One curious result of the experiments made by Dr. Watson and his friends, and which they themselves probably did not expect, was the breaking out of a sort of public frenzy for making like trials, but after the most childish fashion. Everybody who had, or thought he had, the least tincture of science in him, procured a long glass tube, and went on rubbing it assiduously with his handkerchief, sitting in dark rooms and cellars, so as to be better able to watch the first appearance of the 'electric fire.' Ladies and gentlemen alike went on rubbing, with desperate energy, as if the fate of the world depended on their exertions. They sold 'electrical tubes' in pastry shops; every draper praised his own handkerchiefs as the best for rubbing; and lecturers upon electricity went about through the length and breadth of the land, with glass rods in their hands, delivering wonderful harangues, and trying to explain to gaping multitudes the mysteries of nature as regards electricity. The lecturers even crossed the Atlantic to America, visiting the chief towns, and preaching to large assemblies in places—

> Where blind and naked ignorance
> Delivers brawling judgments, unabashed
> On all things, all day long.

If it was a ludicrous spectacle to see these wandering lecturers, with their glass tubes and pocket-handkerchiefs, the movement nevertheless produced, apparently quite by accident, a striking result. It occurred through one of the peripatetic preachers of electric revelation coming face to face with Benjamin Franklin, a printer established at Philadelphia, when on a visit to his native town of Boston, Massachusetts, North America.

There are, in the records of scientific discovery, few figures so interesting, because so full of marked individualism, as that of Benjamin Franklin. He was not a man of genius, in the accepted sense of the word; nor was he even a man of high talents. But he was nevertheless a decidedly great man, his greatness consisting in the largest development of that undefined faculty known as common sense. Benjamin Franklin was the very ideal of a 'practical' man, that is, a man valuing thoughts only as leading to actions, and new ideas only as the road to visible results. The success of his career in life was but an illustration of his thoroughly practical character. Born at Boston in January 1706, the son of a tallow-chandler and soap-boiler, he was destined by his parents to follow the same trade, but not relishing the melting pots, he got apprenticed to an elder brother, a printer at Boston. Harsh treatment drove him away from this place before the terms of his apprenticeship were over, and with scarcely a penny in his pocket, and the experience of only seventeen years in his brain, he made his way to Philadelphia. A year after, when eighteen years of age, he was induced to sail for England, and was fortunate enough to find employment as a compositor in a printing office in London, but so poor as to be compelled to take a lodging for eighteen-pence a week. However, his self-reliance never deserted him; he managed to go unscathed through all the perils of poverty and friendlessness in a great city, and after a few years went back to Philadelphia, with a small stock of money and a wealth of experience. He now set up as a master printer, and gradually, though by very slow degrees and ceaseless toil, devoted to multifarious objects, rose into prosperity. For upwards of twenty years, from 1728 to 1748, he was the most energetic and active man of business in Philadelphia. He was not only a printer, but an author, an editor of newspapers, a compiler of almanacks, a publisher, a bookseller, a bookbinder, and a stationer. He made lamp-black and ink; he dealt in rags; he sold soap and geese feathers; also, as he frequently made known to his fellow-townsmen in printed notices, he had always in stock 'very good sack at six shillings a gallon.' To dispose of the numerous articles in his store he invented the art of advertising, unknown before him at Philadelphia. All the inquiring minds of the 'Quaker City' assembled regularly in the shop of Benjamin Franklin, 'the new printing office near the market,' which came to form the centre of intelligence, and the source from which all public movements went forth.

The reward for all this activity was that at the end of twenty years Benjamin Franklin had accumulated a handsome fortune, his average income amounting to over two thousand pounds sterling a year: then considered a very large sum, and of probably three times the purchase value it would possess at the present day. With increasing wealth, the active printer, happy in all his family relations, thought himself justified to seek a little occasional leisure, which he found chiefly in visits to Boston, his native town. It was on one of these visits, made in the summer of 1746, that he went with a friend to a lecture-hall, scarcely knowing what was to be the intellectual entertainment prepared for him. It proved a discourse, with illustrative experiments, upon electricity, by a Dr. Spence, duly armed with a three-feet glass rod and silk pocket-handkerchief. Benjamin Franklin was not merely interested: he was startled. It was to him, as he afterwards declared to one of his friends, the opening of a new world.

Perhaps the subject which attracted so suddenly the attention of Benjamin Franklin might have escaped it again, in the pursuit of his many vocations, but for another accidental circumstance. It so happened that, immediately after his return to Philadelphia, there came a parcel of books from England, accompanied by a present in the shape of an electrical tube. The sender of it was the London agent of the Library Company of Philadelphia, Mr. Peter Collinson, a member of the Royal Society, and as such sharing the general interest in the electrical experiments of Dr. Watson. The tube, which was accompanied with full directions for its use, was no sooner unpacked, than Franklin seized it eagerly and began experimenting, at the same time inspiring the most sanguine of his friends to follow his example. Glass tubes, made similar to the one sent from London, were soon procured from a local manufacturer, and then began a general rubbing. 'I never before,' Franklin wrote, early in 1747, 'was engaged in any study that so totally engrossed my attention and my time as this has lately done; for what with making experiments when I can be alone, and repeating them to my friends and acquaintances, who, from the novelty of the thing, come continually in crowds to see them, I have, during some months past, had little leisure for anything else.' To the greater number of those friends and acquaintances who came flocking in crowds to the shop of the Philadelphia printer, the electric tube was, probably, only looked upon as a new toy; but it was vastly different as regarded himself. His keen practical eye seemed to discern at once that the manifestations of the mysterious force on which he was experimenting contained the germ of something that might be utilised by men, or brought into obedience to the human will.

It is not very clear from the published correspondence of Franklin what his earliest views on the subject were, but there are many indications that he conceived for a while that the 'electric fire' might be employed in arts and manufactures. In his usual humorous style he spoke of these utilitarian aims of his in a letter to Mr. Peter Collinson, written in the early summer of 1747 : ' Chagrined a little that we have not been able to produce hitherto anything in the way of use to mankind,' he wrote, ' and the hot weather coming on, when electrical experiments are not so agreeable, it is proposed to put an end to them, for this season, in a party of pleasure on the banks of the Schuylkill. Spirits at this party are to be fired by a spark sent from side to side through the river, without any other conductor than the water: an experiment which we some time since performed, to the amazement of many. Then a turkey is to be killed for our dinner by the electrical shock, and roasted by the electrical jack, before a fire kindled by the electrified bottle, when the healths of all the famous electricians in England, Holland, France, and Germany are to be drunk in electrified bumpers, under the discharge of guns from the electrical battery.'

The longer he experimented, the more fascinated grew Benjamin Franklin with his study of the phenomena of electricity. In order to be able to devote himself completely to his darling science, he sold his printing and publishing business in the year 1748, and went to live in a suburb of Philadelphia, not far from the banks of the Delaware. At the same time he purchased a complete set of electrical apparatus, the best that had yet been manufactured, which had been brought over from Europe by the same Dr. Spence who had given him his first ideas about electricity at Boston. With these more perfect means he now continued his investigations, arriving before long at results that formed an epoch in the history of electricity.

The results achieved were wholly of a practical kind. With that strong common sense which formed the most marked feature of his character, Benjamin Franklin, at a very early period of his experiments, came to the conclusion that of the actual *nature* of electricity we know nothing, and, in all probability, never can know anything, with our finite senses. But, never losing sight of this starting point, he treated electricity as astronomers do the movement of the heavenly bodies. Of the incomprehensible forces that keep countless worlds in their courses through measureless space, astronomers know no more than the most ignorant of mankind; still they are able to arrive at very accurate calculations concerning the directions followed by stars and planets, and the amount of time consumed in their wanderings through the inconceivable universe. To such astronomical endeavours Franklin limited all his researches, and it was precisely because he so limited them that he achieved greater successes than any other investigator of the phenomena of electricity.

Together with many smaller matters, Benjamin Franklin added three great discoveries to the knowledge of electricity. The first was that the electric fluid—so called for want of a better word to express the action of the mysterious force—will run its course more easily and quickly through sharply pointed metals than in any other way. This had never before been demonstrated, nor, probably, been ascertained. The second great discovery of Franklin was that of positive and negative electricity, or, as he called it for some time, *plus* and *minus*, the latter names being really the most descriptive. Of the actual existence of these two divisions of the great and marvellous agency, now attracting and now repelling each other, much was known previous to Franklin, but he was the first to make them clearly understood, and to bring their effects within reach of calculation. To these two discoveries Benjamin Franklin added a third, the greatest of all. He established the identity between the electric force and lightning, and upon it based one of the noblest inventions of all ages, that of the lightning conductor. And perhaps there never was any invention acknowledged more deeply by mankind. The French Academy expressed it when, on Franklin's entrance, all the members rose, and the President exclaimed ' *Eripuit cœlo fulmen.*'

The identity of the electric force and lightning, vaguely surmised by previous inquirers, and expressed at times in hints, was not only firmly asserted by Benjamin Franklin, but at a comparatively early part of his investigations proved by him in experiments. His broad practical way of looking at facts succeeded in grasping a truth which all the learned men before him, who had busied themselves with electrical experiments, had not been able to lay hold of, simply because they lost themselves in philosophical abstractions. The professors sought the unattainable, and he confined himself strictly to what he considered within reach, and it was thus he gained his end.

The thoroughly matter-of-fact way in which Franklin went to work is strikingly exhibited in his own description as to how he came to the conclusion of the oneness of lightning and electricity. In reply to a friend and correspondent, living in South Carolina, who had asked him how he came to such an 'out-of-the-way idea' as that of the majestic fire from the cloud-capped firmament being exactly the same with the puny gleam from a stick of glass, rubbed with the sleeve of an old coat, Franklin wrote a highly characteristic letter. 'I cannot answer your question better,' he told his friend, 'than by giving you an extract from the minutes I used to keep of the experiments I made. By this extract you will see that the thought was not so much an out-of-the-way one but that it might have occurred to any electrician. The extract, dated November 7, 1749—a date worth remembrance in the history of scientific progress—was as follows in its entirety :—' Electrical fluid agrees with lightning, in these particulars: 1. Giving light. 2. The colour of the light. 3. In the crooked direction of the flame. 4. In the swift motion. 5. In being conducted by metals. 6. In the crack, or noise, of the explosion. 7. The subsisting in water, or ice. 8. In the rending of bodies it passes through. 9. In destroying animals. 10. In melting metals. 11. In firing inflammable substances. 12. The sulphurous smell. The electric fluid is attracted by points, and we do not know whether this property is in lightning. But since they agree in all the particulars wherein we can already compare them, is it not probable that they agree likewise in this? *Let the experiment be made.*'

CHAPTER II.

DISCOVERY OF THE LIGHTNING CONDUCTOR.

WITH that liberality which distinguishes all truly great minds, Benjamin Franklin did not keep his great discoveries to himself, but communicated them to others in the most open-handed manner. Ever since he had commenced his electrical experiments, he had sent the detailed results of them to his London correspondent, Mr. Peter Collinson, for communication to the Royal Society, and he was not even prevented from continuing the labour of writing long letters by the knowledge of the fact that scant notice was taken of them by the Royal Society. The members of this august learned body, with a few honourable exceptions, seemed unable to hide their contempt for what they considered the dabblings in science of a mere tradesman, living in an obscure little town, in a distant colony. Somebody had mentioned in public that this person, of the name of Franklin, was a dealer in rags and goose-feathers, dwelling among money-worshipping Quakers in the City of Brotherly Love: which naturally was productive of great merriment, but detrimental to scientific respect. Thus, although by the influence of Mr. Collinson and some of his friends, the letters from Philadelphia were read before the Royal Society, they met with scarcely any attention, and the members broadly expressed their disdain of them by refusing to allow their insertion in their 'Transactions.' Three whole years elapsed in this way, when at length, in the autumn of 1750, Benjamin Franklin reported to Mr. Collinson his researches on the identity of electricity and lightning, together with his ideas that all damage done by the electric fire descending from the clouds upon the earth might be put a stop to by fixing iron rods, with sharp points, to the summit of buildings, which would thus be protected. He added that he himself intended shortly to verify his conclusions by experiments, but that, in the meanwhile, it would be well if others did the same. Never before, perhaps, was a grand idea thrown out to all the world with more munificence of spirit, and with more entire abnegation of the very thought of self.

Franklin's letter made a great impression upon Mr. Collinson. Anxious to make it public, while persuaded that the Royal Society would give no better reception to it than to the author's previous communications, he hastened to Mr. Edward Cave, proprietor and editor of the 'Gentleman's Magazine,' and asked him to print it in his publication, the most widely read at the time. A man of quick sense, Mr. Cave, too, saw at once the vast importance of Franklin's paper, describing his discovery, and readily offered to print it, but recommended that it should be done in pamphlet form, as likely to make the facts even more extensively known than could be the case in his own Magazine. This having been agreed to, there appeared, early in May 1751, a pamphlet with the name of Benjamin Franklin on the front page, and a preface by Dr. Fothergill, entitled, 'New Experiments and Observations in Electricity, made at Philadelphia, in America.' It was the most important contribution to science published since the appearance, five-and-thirty years before, of Newton's 'Principia.'

Like Newton's book, that of Franklin was not immediately successful—at least not in England. Not appearing under the patronage of the Royal Society, the supposed fountain-head of all legitimate science, it was looked coldly upon by the public and the critics, and it was only after having been greeted with immense applause in France, that at last something like justice was done to it in England. The great success of Franklin's little treatise in France was due, in the first instance, to rather accidental circumstances, but was none the less genuine. By a happy chance a copy of the pamphlet printed by Mr. Cave fell into the hands of the Count de Buffon, the greatest naturalist of the age, and whose pre-eminent position was established not only in France, but throughout the whole of Europe. Himself familiar with the English language, he yet thought that it was necessary to have the book immediately translated into French, and he employed for the purpose Professor Dubourg, a literary man of note, well versed in electrical science. Under such favourable auspices, Franklin's pamphlet, carefully translated, was issued at Paris in the summer of 1751, three or four months after its appearance in London. Its success in France was as immediate as it was great, and the wave of it spread at once over Europe, marked by German, Italian, and Latin translations of the 'New Experiments.' For a considerable time nothing was talked of among the upper classes of France but the discoveries in science of the unknown Philadelphia printer, and the king, Louis XV., following the fashion of the day, ordered a course of the electrical experiments, described by Franklin, to be performed before him at St. Germain, in the presence of the whole court.

A rather ludicrous incident, and which gave rise to a great deal of scientific tournamenting, added to the celebrity of Franklin's little book on the continent of Europe. The greatest of French electricians, Abbé Nollet, a man of acknowledged merit, but inordinately vain, was mystified in believing that the pamphlet which caused such an immense stir at court and among the public was not the production of the obscure man Franklin of Philadelphia, but got up among his enemies in England and France, to rob him of his reputation. With this belief fixed in his mind, he sat down at his desk to write a series of letters intended to demolish the man of Philadelphia, and proving, entirely to his own satisfaction, first, that Franklin did not exist at all; secondly, that he had no right to exist; and thirdly, that all his pretended discoveries were mere dreams. Not long after the publication of his letters, the wrathful Abbé received undoubted proofs from America that at Philadelphia there was a man called Franklin, who himself mildly asserted his right not only to live, but to make experiments in electricity. Poor Abbé Nollet felt his humiliation all the more keenly as holding the post of preceptor in Natural Philosophy to the royal family of France, and he had to suffer from a 'burst of inextinguishable laughter' at one of his appearances at court.

If Count de Buffon did great service to electrical science by getting Franklin's pamphlet translated into French, he did still more by instigating a series of experiments tending to verify the great theory put forward in the pamphlet, that lightning could be drawn from the clouds by means of pointed iron rods. By his prompting, several gentlemen interested in scientific pursuits engaged upon trials to this effect, among them two persons of note, M. Dalibard and M. de Lor. The first-named had the good fortune to be successful, and thereby to hand his name down to posterity. A wealthy man of science, M. Dalibard was in the habit of living, during a part of the year, in a handsome country house situated at Marly-la-Ville, about eighteen miles from Paris, on the road to Pontoise. Marly-la-Ville stands on a high plain, some four hundred feet above the sea-level, and the residence of M. Dalibard being situated on the most elevated part of the ground, it formed an excellent place for experiments, and was chosen as such by Count de Buffon. The garden near the house was selected as the best ground for the experiments. A wooden scaffolding was built up to hold in its midst an iron rod, eighty feet long, and slightly over an inch in diameter. On the top of the rod was fastened a piece of polished steel, sharply pointed, and bronzed to prevent rust. The iron rod entered, five feet from the ground, into another thinner one, running horizontally towards an electrical apparatus, fastened to a table in a kind of sentry-box, erected on purpose for observations. It was

M. Dalibard's intention to make the experiments himself; but almost immediately after the structure in his garden had been completed, he was called by business to Paris, and left the whole in charge of one of his servants, an old soldier, formerly in the French dragoons, Coiffier by name. With true military spirit, Coiffier thought that he ought to spend the greater part of his time in the sentry-box in his master's garden, and there he sat in the afternoon of Wednesday, May 10, 1752, when a violent thunderstorm drifted over the plain of Marly. Sufficiently instructed by his master what to do under the circumstances, he touched the electrical apparatus with a key, silk-bound at the handle, and to his extreme surprise, sees a flame bursting forth. He touches another time, and there is a second flame bursting forth, stronger than before. Then the old dragoon rushes from his sentry-box—most famous private dragoon that ever lived, born to the high honour of being the first man that ever drew lightning from heaven.

It was not fear that drove the worthy servant of M. Dalibard from his post, but a far better motive. He judged, with the prudence of an old soldier, that the astounding things he had seen required witnesses, in order that his master might not think him an inventor of fairy tales. Accordingly, he hurried to the house of the prior of Marly, M. Raulet, who lived close by, and asked him to behold the marvel of marvels. The prior hesitated not for a moment to go, and, entering the sentry-box, he also drew sparks from the electrical machine. Others of the inhabitants of Marly-la-Ville, seeing the prior run, followed in his wake, notwithstanding the rain was pouring down in streams, and terror was struck among all of them in witnessing the dreaded lightning creep down, serpent-like, but bereft of all its terrors, into the sentry-box, in the centre of which stood the now exulting old dragoon. As soon as the storm was over, the prior insisted upon Coiffier at once saddling a horse, and riding full speed to Paris, to acquaint his master with the great news that lightning had been drawn from the skies by his apparatus at the blessed village of Marly-la-Ville. The obedient dragoon did as advised, and three days after, on May 13, 1752, M. Dalibard startled all the members of the Académie des Sciences of Paris, convoked together in haste, by reading to them a full report of what had taken place in the first great experiment for ascertaining the truth of the suggestions of Benjamin Franklin.

All Europe soon rang with the report of the marvellous discovery verified at Marly-la-Ville. But before the news of the experiment made at the village near Paris had reached America, Benjamin Franklin had made another, which, if not more conclusive, was at least more original. Ever since he had arrived at his great conclusion regarding the sameness of electricity and lightning, and the possibility of conducting the latter to the ground harmlessly, by means of pointed rods, the discerning citizen of Philadelphia had tried hard to find some means for putting his ideas to a practical test, but met with apparently insurmountable difficulties. His first plan was to set up simply a tall iron rod near his house; but he abandoned this on ascertaining, by measurement, that nearly all stormclouds passed over Philadelphia, which was situated in a plain, at a height of several hundred feet. In his then state of knowledge, he fancied that it was impossible for him to reach the clouds in this manner. He next resolved to await the building of an intended steeple for the principal ecclesiastical edifice, and highest building of Philadelphia, Christ Church. At that time not a steeple pierced the sky in all the extent of the ' Quaker city;' nor was there a single one in the whole State of Pennsylvania. But though Franklin made immense efforts to get the steeple erected, starting a lottery for the purpose, and subscribing largely to the funds, the work made little or no progress, many of the principal inhabitants of the city being, from their religious opinions, averse to the project. At last, getting impatient, Franklin's ingenuity hit upon the simplest of all means for verifying his great discovery.

One day he saw a boy flying a kite, and the thought instantly occurred to him that here was the straight road from the earth to the thunderclouds. Accordingly, he at once set to make a kite for his intended experiments; but fearing he would incur the ridicule of his sober fellow-citizens in engaging in what might seem to them a childish undertaking, he kept the whole matter a profound secret. The kite he made was not distinguished from those used by boys except of being made of silk instead of paper so as to be able to stand the wet. Franklin took an ordinary silk pocket-handkerchief, and fastened it over a cross made of two light strips of cedar, by simply tying the four corners of the handkerchief to the ends of the sticks. He next fastened a thin iron wire, a foot long, to the top of the kite, and having provided it with a loop and tail, attaching to the former a roll of twine, all was ready for the experiment. Watching the skies diligently, he saw a dark thundercloud coming up over Philadelphia late in the evening of July 4, 1752, and at once sallied forth from his house, situated at the corner of Race and Eight streets, into a neighbouring field. There was nobody with him but his eldest son, a lad of about twenty; and, in order to get protection against the heavy downpour, as well as to hide from the gaze of passers-by, the two sought shelter under an old cow-shed. Very likely, had they been seen here at the time, the philosopher and son might have been taken for two escaped lunatics, seeking so propitious an occasion as a thunderstorm to fly their darling kite. Perhaps Franklin too felt a little foolish, for he was about relinquishing his experiment after several flashes of lightning which had not in the least disturbed his kite, when a cloud darker than the previous one came rolling up. All on a sudden, Franklin felt a smart shock, and saw a spark flashing before his eyes. He had fastened the twine holding his kite to a silk ribbon which he held in his hand, joining twine and silk by a large key, attached to a Leyden jar. The latter at once became heavily charged, and as shock followed upon shock, and flash upon flash, there vanished all doubt from Franklin's mind as to the absolute truth of the grand discovery he had made. It may be imagined with what inward satisfaction the great citizen of Philadelphia drew in his kite, and crept out from under the cow-shed, when the storm was over, and went home exultingly, the happiest of philosophers.

The experiment of Benjamin Franklin in drawing, as he thought, the electricity of stormclouds to the ground by his kite, and thereby demonstrating the necessity for the establishment of lightning conductors, for the protection of persons and buildings, was accepted as thoroughly satisfactory by the whole scientific world of Europe at that time. Franklin was wrong, however, in supposing that the lightning had really passed along his kite-string from the clouds to the earth, for, had this been the case, he would undoubtedly have been killed. What he witnessed was merely the inductive action of the thundercloud on the kite and string. There had been some doubts in respect to the experiment made, at the suggestion of Franklin's pamphlet, at Marly-la-Ville, since all the witnesses were inexperienced persons, entirely unacquainted with the phenomena of electricity; but there could be none whatever as regarded that tried by the originator himself, and pronounced satisfactory by him. The fame of the wonderful discovery spread with extraordinary swiftness through the civilized world. Praises and congratulations flowed in upon the hitherto obscure citizen of Philadelphia from all sides. The king of France sent him a letter, full of compliments; the Royal Society of London voted him their gold medal, modestly claiming a share in his work; and nearly all the scientific bodies of France, Germany, and Italy elected him an honorary member. But the praise of which Franklin had most reason to be proud came from the great philosopher Immanuel Kant. The sage of Königsberg grandly called him the modern Prometheus, bringing fire from heaven.

CHAPTER III.

EARLY EXPERIMENTS WITH LIGHTNING CONDUCTORS.

THE first actual lightning conductor ever constructed was set up by Benjamin Franklin himself, at his house in Philadelphia. Its main object was to protect the house against the effects of thunderstorms; still experiments were so dear to the heart of the great discoverer, that he could not help making trials even with things devoted to other uses. It was in the summer of 1752 that Franklin erected over his house a lightning conductor, made entirely of iron, but with a sharp steel point on the top, the latter projecting seven or eight feet above the roof, while the end was above five feet in the ground. Curious to know whenever an electrical stream was passing through the conductor, he attached to it an ingenious contrivance, by means of which through an electric spark two bells were set in movement as soon as this took place, the greater or lesser noise from them corresponding with the strength of the electrical current. With the aid of this device Franklin was enabled to observe some curious phenomena, which at first puzzled him not a little. 'I found the bells rang sometimes,' he informed a friend, 'when there was no lightning or thunder, but only a dark cloud over the rod; that sometimes, after a flash of lightning, they would suddenly stop, and at other times, when they had not rung before, they would, after a flash, suddenly begin to ring; that the electricity was sometimes very faint, so that when a small spark was obtained, another could not be got for some time after. At other times, the sparks would follow extremely quickly; and once I had a continual stream from bell to bell, the size of a crow-quill. Even during the same gust there were considerable variations.' By continued watching, Franklin came to make the discovery that the fluctuations in the electrical current were owing to changes and interchanges, in atmosphere and earth, of positive and negative electricity. He held at first that thunder-clouds are usually in a negative state of electricity, but afterwards discovered that they varied from negative to positive during the same storm.

Notwithstanding the unbounded praises bestowed upon Benjamin Franklin for the great discovery of the lightning conductor, the actual adaptation of it spread with extreme slowness. It was in the country of its origin that it was brought into public use, all the countries of Europe lagging far behind. But even in the Northern States of America, though inhabited by a highly intelligent race, there were great difficulties to overcome. The ministers of religion at first seemed to think that the iron rods were not altogether free from the suspicion of infidelity. Franklin himself had the reputation of being a free-thinker, and indeed never hid from others the fact of his being accustomed to examine all matters by the light of his own reason, and to believe nothing that he could not understand. Perhaps on the same ground many of the New England ministers did not believe in lightning conductors. They could not understand them. A heavy shock of earthquake was felt throughout Massachusetts in the summer of 1755, whereupon a Boston clergyman instantly came forward, denouncing in eloquent strains the erection of a number of lightning conductors which had taken place. The high iron rods, he gravely maintained, had been the cause of the earthquake, by drawing vast masses of electricity from the atmosphere into the ground. A distinguished friend of Franklin, Professor Winthrop, of Harvard College, thought it necessary to come forward and defend lightning conductors against the accusation of accumulating electricity, but without convincing the plaintiff. A different charge, still more serious in the eyes of pious people, had been made against lightning conductors some years before. Another Boston clergyman, coming forward in 1770, opposed the use of Franklin's iron rods on the ground that, as the lightning was one of the acknowledged means of punishing the sins of mankind, and of warning them from the commission of acts of wickedness, it was impious 'to prevent the execution of the wrath of heaven.' To this gentleman also Professor Winthrop deemed it requisite to reply. Franklin himself remained silent, wrapping himself in the mantle of the sage. But he allowed his friend Ebenezer Kinnersley, of Philadelphia, who went travelling, by his wish and partly at his cost, through the principal towns and villages of the New England States, to explain to the people the uses and advantages of lightning conductors, to preface all his lectures by the announcement that the erection of iron rods to protect houses from the effects of thunderstorms was not an act 'chargeable with presumption, nor inconsistent with any of the principles either of natural or revealed religion.'

In the gradual spread of lightning conductors through the British colonies of North America, Franklin himself took the leading part. He employed all his leisure time, engrossed though it was more and more by political affairs, in which he was destined to take a world-famous part, in going from one part of the country to another, advocating the use of conductors, advising as to the best mode of their construction, and, whenever he could, examining into the effects of strokes of lightning upon buildings. How minute he was in these inspections, and how practical in the conclusions he almost invariably drew from them, Franklin gives proof in one of his letters addressed to his friend Collinson in London. He tells him that he inspected the church of Newbury, in Massachusetts, which had been struck by lightning, and traced, foot by foot and inch by inch, the road which the electric current had taken, creating great havoc and destruction. 'The steeple,' he says, 'was a square tower of wood, reaching seventy feet up from the ground to the place where the bell hung, over which rose a taper spire, of wood likewise, reaching seventy feet higher, to the vane of the weather-cock. Near the bell was fixed an iron hammer to strike the hours; and from the tail of the hammer a wire went down through a small gimlet-hole in the floor the bell stood upon; then horizontally under and near the plastered ceiling of that second floor, till it came to a wall; and then down by the side of this wall to a clock which stood about twenty feet below the bell. The wire was not bigger than a common knitting-needle.' It surprised Franklin that 'the lightning passed between the hammer and the clock in this wire, without hurting either of the floors, or having any effect upon them, except making the gimlet-holes, through which the wire passed, a little bigger, and without hurting the wall or any part of the building.' The inference he drew from this was, that even a comparatively thin mass of metal would give passage to a powerful electric stream. 'The quantity of lightning that passed through the steeple,' he informed his correspondent, 'must have been very great, as shown by its effects on the lofty spire above the bell, and on the square tower below the end of the clock pendulum; and yet, great as this quantity was, it was conducted by a small wire and a clock pendulum, without the least damage to the building as far as they extended.'

Besides travelling and employing lecturers, to make the advantages of lightning conductors known, Franklin found means of doing so in an annual publication he had started in the year 1732, known as 'Poor Richard.' This almanac, humorous in form but very serious in substance, which had acquired an enormous circulation, proved in the end the most powerful instrument for spreading information on the great subject dear, above all others, to Franklin's heart, and leading his countrymen to adopt, before all other nations, the wonderful metal rod, protective against 'the wrath of heaven.' In several of the editions of the almanac, notably the 'Poor Richard' for the year 1758, Franklin drew attention to his lightning conductors in simple advertisements, drawn up in a spirit of absolutely touching modesty and self-abnegation. Not seeking the slightest reward for himself, nor even mentioning his name, he only sought to benefit others by instructing them how to get protection against the dangers of lightning. 'It has pleased God,' ran the advertisement in the almanac, ' in His goodness to mankind, at length to discover to them the means of securing their habitations and other buildings from mischief by thunder and lightning. The method is this :—Provide a small iron rod, which may be made of the rod-iron used by nailers, but of such a length that, one end being three or four feet in the moist ground, the other may be six or eight feet above the highest part of the building. To the upper end of the rod fasten about a foot of brass wire, the size of a common knitting-needle, sharpened to a fine point; the rod may be secured on the house by a few small staples. If the house or barn be long, there may be a rod and point at each end, and a middling wire along the ridge from one to the other. A house thus furnished will not be damaged by lightning, it being attracted by the points and passing through the metal into the ground without hurting anything. Vessels also, having a sharp-pointed rod fixed on the top of their masts, with a wire from the foot of the rod reaching down round one of the shrouds to the water, will not be hurt by lightning.' Franklin had occasion subsequently greatly to modify the advice here given. He early discovered his error of lightning being ' attracted by the points ;' and also found that his recommendation to people to construct their own lightning conductors only led to grievous calamities. There came reports from all sides of houses having been severely damaged by lightning notwithstanding having conductors, and close investigation soon showed that in every instance the apparatus was defective, having been erected by unskilful hands, either the owners themselves, or a set of wandering impostors, who soon made themselves notorious as 'lightning-rod men.'

Having improved in various ways the lightning conductor

set up experimentally over his own house, Franklin erected a second one, of larger dimensions, to protect the residence of one of his friends, Mr. West, a wealthy merchant of Philadelphia. The apparatus, constructed entirely under the supervision of Franklin, consisted of an iron rod half an inch in diameter throughout its length, and ending at the bottom in a thick iron stake, driven four or five feet into the ground. The top of the conductor, rising nine feet above the central stack of chimneys, was formed by a brass wire ten inches in length, tapering off in a sharp point. Franklin considered the brass wire, which was screwed and soldered inside the iron rod, a great improvement upon simple iron, having discovered brass, as well as copper, to be better conductors of electricity. The result justified his expectations. Not many months after the lightning conductor had been erected over the mansion of Mr. West, a thunderstorm more severe than had been experienced for many years broke over Philadelphia. Vivid flashes of lightning followed each other incessantly, one of them striking, visible to all beholders, the house of Mr. West, touching the point of the conductor on the roof, and appearing again on its base in a thin sheet of flame. Naturally, Franklin was delighted at this first notable result of his grand discovery, and lost no time in examining the traces of the lightning over his conductor. He found that the sharp metal point at the upper end had been melted, and the small brass wire reduced from ten to seven and a half inches, with its top very blunt. The thinnest part of the wire, he saw at once, had disappeared in smoke, while the portion below it, a little thicker, had simply been liquefied, sinking down while in a fluid state, and forming a rough irregular cap, lower on one side than on the other. This was a highly interesting test, showing that the wire on the summit of the conductor must not be made too thin, so as to be liable to be burnt. But still more interesting to Franklin was the investigation of the report, confirmed on all sides, that a sheet of flame had been seen at the base of the conductor, where it was connected with the earth. He at once suspected that the earth at the point, and down to the end of the metal rod, had been very dry, and such indeed was the case. Hence he arrived at the conclusion that all conductors should go deep enough into the earth to find sufficient moisture quickly to dissipate the electric fluid. All subsequent experience, down to the present day, has proved that the inference of the practical philosopher of Philadelphia was as sound in this respect as in the rest of his ever clear and lucid judgments.

Like most other inventions and discoveries, that of the lightning conductor was destined not to be without its early martyrs. Among the many searchers in the science of electricity on the continent of Europe who had eagerly seized the ideas of Benjamin Franklin, and entered enthusiastically upon the experiments recommended by him, was Professor George Wilhelm Richmann, of St. Petersburg. He had conceived some theories of his own regarding electrical discharges, and constructed for experimental purposes an apparatus which he called the 'gnomon,' one of the uses of which was to measure the comparative strength of electrical currents. The instrument consisted of a tube of metal, terminating in a small glass vessel, into which, for some unknown reason, he put a quantity of brass filings. Attached to the tube of metal, at its top, was a chain, so arranged as to be easily attached or detached from it, and this was fastened to an iron rod going to the roof, in the form of a lightning conductor, as prescribed by Franklin. It seems to have been the notion of the professor that he might lead the electrical current from the clouds down into his 'gnomon' bottle, there to measure its strength; though it is difficult to conceive how a man acquainted with the manifestations of the mystic force with which he was experimenting, and knowing its powerful effects, should not have perceived the extreme danger of thus leading it into a non-conducting element. However, the enthusiastic man, evidently blind to all consequences, set out on his course of experiments. A violent thunderstorm coming over St. Petersburg on August 6, 1753, Professor Richmann hurried to his 'gnomon,' attached the chain to the phial, and then stood to watch the effect, with not more than a foot and a half distance between his head and the glass tube. Near him, but further behind, stood a friend, M. Solokow, who was going to make a drawing of the electrical apparatus. All on a sudden, there came a terrible flash of lightning, described as 'a ball of fire' by M. Solokow, down from the skies, falling upon the 'gnomon' and springing from thence upon Professor Richmann, laid the latter dead on the floor, and his companion senseless.

When the body of the unfortunate professor came to be examined, it was found that the electric current had passed right through him, entering at the forehead, and coming out at the sole of the left foot, both places being distinctly marked by red spots and small perforations, like those of a needle. There were no other marks of injury visible, either inwardly or outwardly, except a number of red and blue spots over the back and shoulders, which grew larger the day after, and seemed to bring with them symptoms of rapid decay. Some of the medical men attending the 'post mortem' examination were most desirous to enter into further observations, so as to ascertain, if possible, the actual cause which produced death by a stroke of lightning, but they had no opportunity. When they returned to the professor's house, the second day after his death, the body was already so far decomposed as to be unrecognisable, and it was with difficulty that the remains of the first martyr of applied electricity could be got into a coffin and carried to their last resting-place.

The appalling death of Professor Richmann produced an enormous commotion, far beyond what might be expected from a similar event, throughout the learned world of Europe. In France especially the occurrence created the deepest impression, mingled with admiration of what was called the 'glorious death' of the St. Petersburg professor, and more than one student of electrical science boldly declared his determination to become a martyr in the same noble cause. But reflection, probably, brought better counsel, for, as it happened, there were no more contributions, for the time being, to the roll of martyrs.

CHAPTER IV

GRADUAL SPREAD OF LIGHTNING CONDUCTORS IN EUROPE.

In singular contrast with the burst of applause with which the whole scientific world of Europe received the great discovery of Benjamin Franklin, was the extreme slowness of the actual introduction into Europe of lightning conductors. The opposition they met with in Franklin's own country was trifling to that which they encountered in the principal states of Europe, more particularly in England and France. It was natural, perhaps, that the lower classes—ultra-conservative, through the mere effect of ignorance, in every country in the world—should see danger in the setting-up of iron rods which, as they were told, drew lightning from the skies; and it was, perhaps, equally natural that religious fanatics should regard them with extreme suspicion, as removing one of their imagined instruments of heaven for punishing sinful mortals. Both these classes, the untaught multitude and the bigoted zealots, opposed in Europe, as they did in America, the establishment of lightning conductors; but to the strength of these parties was unexpectedly added a third in a not numerous but powerful section of learned literary men. They were chiefly French, but had many adherents in England, as well as in Germany, the *savans* of both countries looking then upon France as the seat of all science, and indeed human knowledge.

The opposition raised against lightning conductors in France was entirely personal, its origin being due to the wounded vanity of a very estimable but likewise a very weak man, the already mentioned Abbé Nollet. Born in 1700, the Abbé had very early in life gained renown for his scientific researches, and after a while devoted much of his time to electrical experiments, in conjunction with two other celebrated men, Dufay and De Réaumur. When the report of Franklin's discoveries arrived in Europe, the Abbé Nollet was generally looked upon as the greatest of living 'electricians,' and the general homage paid to him having roused his self-esteem to an inordinate degree, he got fiercely irritated that another man, a previously quite unknown person, in a distant land, should have dared to snatch from him his scientific laurels. Accordingly, he used all his influence among the public, in the scientific world, and at the French court, where he held a high position as tutor of the King's children, not only to depreciate Franklin's lightning conductors, but to set them down as something like an imposture. In various treatises and articles published in learned papers, Abbé Nollet sought to prove that the person called Benjamin Franklin—in whose very existence he formerly refused to believe, but which he now grudgingly acknowledged—was an individual unacquainted even with the first principles of the science of electricity, and that his proposal for protecting houses against lightning was so absurd as not to be worth engaging the attention of any thinking man. More than this, he argued that the proposed lightning conductors were not only inefficacious, but positively dangerous. By thus joining in the vulgar cry of lightning being, so to speak, sucked from the clouds by Franklin's conductors, the learned Abbé had the satisfaction of retarding their introduction in his own, as well as other European countries, for a number of years.

In France itself the thus awakened resistance to the setting-up of lightning conductors was strikingly shown by an incident which occurred at the town of St. Omer, not far from Calais. A manufacturer settled here, who had been in America, and there learnt to appreciate the usefulness of Franklin's lightning conductors, had one made for his own house, and quietly fixed it to wall and roof. But the populace no sooner heard of it when there arose a public disturbance, and the iron rod was torn down by force. So far from repressing the rioters, the municipality of St. Omer, acting under priestly influence, forbade the manufacturer to erect another lightning conductor, on the ground that it was 'against law and religion.' Thereupon the bold manufacturer, a man of English descent, to try his right, appealed to the tribunals, and the judges at last, after protracted pleadings, not being able to discover any statutes against the fastening of metal rods to buildings, declared that the thing might be done, but with precautions. The lawyer who pleaded the case of the lightning conductors before the French tribunals at this momentous period was a very young man, quite unknown to fame at the time, but destined for a superabundance of it. His name was Robespierre.

Perhaps the violent opposition which the erection of lightning conductors—or 'Franklin rods,' as they were often called—met almost everywhere, would have proved more effective than it ultimately turned out, had not the great discoverer himself showed admirable temper in meeting his enemies, thus pouring oil upon the stormy waters. His calmness and confidence is admirably shown in a letter, dated July 2, 1768, addressed to Professor John Winthrop, of Cambridge, in answer to one in which astonishment was expressed at the 'force of prejudice, even in an age of so much knowledge and free inquiry,' of not placing lightning conductors upon all elevated buildings. Franklin—or he must now be called Dr. Franklin, having received the degrees of LL. D. and D. C. L. from the universities of St. Andrew's, Edinburgh, and Oxford—was residing in England at the time, as agent of the people of Pennsylvania. He was thoroughly acquainted with the state of public feeling, yet so far from being angry, smiled down upon it like a true philosopher. 'It is perhaps not so extraordinary,' he wrote to his friend, 'that unlearned men, such as commonly compose our church vestries, should not yet be acquainted with, and sensible of, the benefits of metal conductors in averting the stroke of lightning, and preserving our houses from its violent effects, or that they should still be prejudiced against the use of such conductors, when we see how long even philosophers, men of science and of great ingenuity, can hold out against the evidence of new knowledge that does not square with their preconceptions; and how long men can retain a practice that is conformable to their prejudices, and expect a benefit from such practice, though constant experience shows its inutility. A late piece of the Abbé Nollet, printed last year in the Memoirs of the French Academy of Sciences, affords strong instances of this; for though the very relations he gives of the effects of lightning in several churches and other buildings show clearly that it was conducted from one part to another by wires, gildings, and other pieces of metal that were *within*, or connected with the building, yet in the same paper he objects to the providing of metallic conductors *without* the building, as useless or dangerous. He cautions people not to ring the church bells during a thunderstorm, lest the lightning, in its way to the earth, should be conducted down to them by the bell ropes, which are but bad conductors; and yet he is against fixing metal rods on the outside of the steeple, which are known to be much better conductors, and through which lightning would certainly choose to pass, rather than through dry hemp. And though, for a thousand years past, church bells have been solemnly consecrated by the Romish Church, in expectation that the sound of such blessed bells would drive away thunderstorms, and secure buildings from the stroke of lightning; and, during so long a period, it has not been found by experience, that places within the reach of such blessed sound are safer than others where it is never heard, but that, on the contrary, the lightning seems to strike steeples by choice, and at the very time the bells are ringing, yet still they continue to bless the new bells, and jangle the old ones whenever it thunders.'

'One would think,' continues Dr. Franklin, with exquisite

humour, 'that it was now time to try some other trick. Ours is recommended, whatever the able French philosopher may say to the contrary, by more than twelve years' experience, during which, among the great number of houses furnished with iron rods in North America, not one so guarded has been materially hurt by lightning, and many have been evidently preserved by their means; while a number of houses, churches, barns, ships, &c., in different places, unprovided with rods, have been struck and greatly damaged, demolished, or burnt. Probably, the vestries of English churches are not generally well acquainted with these facts; otherwise, since as good Protestants they have no faith in the blessing of bells, they would be less excusable in not providing this other security for their respective churches, and for the good people that may happen to be assembled in them during a tempest, especially as these buildings, from their greater height, are more exposed to the stroke of lightning than our common dwellings.'

While Franklin thus wrote of 'the great number of houses furnished with iron rods in North America,' there was not a single public building so protected in England. Several private persons had adopted them for their houses, following the example of Dr. William Watson—subsequently Sir William—vice-president of the Royal Society, who had been the first to set up a lightning conductor in England, erecting one over his cottage at Payneshill, near London, in 1762. But notwithstanding the evident utility of the 'Franklin rods,' they were refused where they were most wanted—for larger buildings, and particularly for churches. The 'unlearned men, such as commonly compose our church vestries,' openly declared against them, and among the clergy there was a steady, if often silent, antagonism to their introduction. The first movement towards its being upset was given by an occurrence which caused much commotion, and gave rise to a vast amount of discussion. On Sunday, June 18, 1764, a few minutes before three in the afternoon, the splendid steeple of St. Bride's Church, in the city of London, one of the architectural monuments of Sir Christopher Wren, was struck by lightning, the flash being intensely vivid, blinding several people. The damage done was so serious that about ninety feet of the steeple had to be taken down entirely, while great and expensive repairs were required for the rest. Dr. Watson, as the first introducer, so one of the chief promoters of Franklin's invention in England, took this opportunity of publishing in the 'Philosophical Transactions' a detailed account of the effects of lightning upon St. Bride's steeple, explaining the potency of conductors in the very action of the electric force. He showed how the lightning first struck the metallic weathercock at the top of the steeple, and ran down, without injuring anything, the large iron bars by which it was supported. At the bottom of the bars, the electric force shattered a number of huge stones into fragments, to make its way to some other pieces of iron, inserted into the walls to give them strength. So it went on till there were no more metals, when havoc and destruction became the greatest. Thus, as Dr. Watson conclusively proved, the beautiful steeple of St. Bride was wilfully made over to ruin for want of a few hundred yards of iron, or other metal, which would lead the electric force harmlessly from the weathercock on the summit into the earth. He finished by telling in the plainest terms, to all on whom devolved the duty of taking care of churches, that it was neglectful, even to criminality, not to protect them by conductors against the always imminent danger of being struck by lightning.

The lay-sermon of Dr. Watson, deeply impressive by the power of the indisputable facts on which it was based, had a considerable effect in rousing public opinion, finding its way even into the dull ears of 'such as commonly compose church vestries.' Among the most important results was a step taken, after long and solemn deliberations, extending over several years, by the Dean and Chapter of St. Paul's. They made an application to the Royal Society, asking for advice as to the best means of protecting the great cathedral,

Sir Christopher Wren's noblest creation, against the perils of lightning. The application was made on March 22, 1769, as recorded under that date in the 'Gentleman's Magazine.' 'A letter from the Dean and Chapter of St. Paul's,' it was stated, 'was read at the Royal Society, requesting the direction of that learned body for the sudden effects of lightning. It was referred to a committee consisting of Dr. Franklyn (*sic*), Dr. Watson, Mr. Canton, Mr. Edward Delaval, and Mr. Wilson, who, after having examined the building, are to report their opinion.' The committee thus nominated embraced all the most eminent men of the day who had studied the phenomena of electricity, and in the order in which they ranked. Next to the great discoverer of the lightning conductor himself, Dr. Watson could claim to stand; and next to him Mr. John Canton, a most painstaking and intelligent worker in the field, inventor of the pith-ball electrometer, and other instruments.

But a curious element of discord pervaded from the first this small conclave of learned men, chosen to decide the not unimportant question as to the best means of providing the cathedral of St. Paul with lightning conductors. That the noble building should be so protected, all were agreed; and it was clearly understood, besides, that if once St. Paul's had lightning conductors, all the other cathedrals and principal churches of England would follow suit. What they differed upon was not this, but the best form of lightning conductors. Franklin's steadfast assertion that points to the elevated rods were not only far preferable to any other form of conductors, but the only really protective ones, was adopted by Dr. Watson and Mr. Canton; but they were opposed by Mr. Wilson, who asserted, with some degree of vehemence, that points were dangerous, and that balls on the summit of the rods afforded infinitely better protection. Standing alone in this view among the eminent members of the committee of the Royal Society, his arguments naturally had no effect, and the recommendation to the Dean and Chapter of St. Paul's was to protect the cathedral by pointed lightning conductors. This was done accordingly. 'Franklin rods' were attached to Wren's splendid structure, worthy to be the introducer of them, on a large scale, in Europe.

The dispute as to pointed conductors, or balls, was by no means brought to a termination by the decision that was come to regarding St. Paul's. Endless pamphlets were published on the subject, and it went so far as to being turned into a political question. As priests scented heresy in the daring attempt to draw lightning from the clouds, so the court faction and ultra-conservatives of England smelt republicanism in the erection of iron rods designed by the representative of the disaffected American colonies. The king was understood to have given his own high opinion entirely against points, and in favour of balls, declaring his preference by ordering a cannon ball of large size to be placed on the top of a conductor erected over the royal palace at Kew. Meeting such high patronage, the 'anti-Franklinians' only sought an occasion to break out into open scientific warfare, and they were not long in finding it. On May 15, 1777, a large public building at Purfleet, on the Thames, serving as a storehouse for war material, was struck and greatly damaged by lightning, although protected by a pointed lightning conductor. Thereupon arose an instant outcry against the system advocated by Dr. Franklin. From much evidence adduced, there could be no doubt that the building at Purfleet had been hurt simply because the conductor was defective in parts, and was besides not laid deep enough into the ground; still this did not stop the clamour raised. Chiefly through the agitation of Mr. Wilson, the members of the Royal Society entered into hot discussions about the respective merits of pointed and round conductors. The feeling of the partisans of the latter side ran so high on this occasion, that Sir John Pringle had to resign the presidency of the Royal Society, which post he had ably filled since 1772, for making himself an advocate of points against balls. When the fever of the learned men had cooled down a little, it was resolved to settle the great question of points *versus* balls by a series of experiments, to be held in the Pantheon, a large building in Oxford Street, dome-like in the interior. The arrangement, in fact, carried out under the direction of Mr. Wilson, leader of the 'ball' party, was to create an artificial thunderstorm—or, as it should properly be called, 'lightning storm '—by means of powerful electrical batteries, to be discharged upon conductors of various forms. His Majesty George III., greatly interested in the subject, and cherishing fond hopes that cannon-balls would carry off the victory in the scientific dispute, as well as in the graver political one with Franklin's countrymen, undertook to pay all the expenses of the Pantheon experiments, and they took place accordingly on an elaborate scale. But though prepared entirely with a view of showing the inefficiency of Dr. Franklin's points, they proved absolutely the contrary. Artificial, like real, lightning clearly showed its preference for a lancet over a ball; it would glide down the former quietly, but fall heavily, mostly with an explosion, upon the latter. However, the question being in reality less a scientific controversy than a dispute arising from the fiery heat of political passions, it was by no means set at rest by the Pantheon trials. 'Franklin rods' were more than ever abhorred by a multitude of persons, learned and unlearned, after the great citizen of Philadelphia had set his hand, on July 4, 1776, to the declaration of independence of the 'United States of America,' and more than a quarter of a century had to elapse, a new generation of men growing up, before there arose clear and unimpassioned views about lightning conductors.

While thus the battle of the rods was being fought in England, it raged no less hotly on the continent of Europe. Here there was religious prejudice alone at work, the political sympathies running in favour of anything coming from America. But priestly animosity by itself proved as strong an obstacle as any other to the erection of lightning conductors. Where it did not exist, they sprang up with rapidity; but wherever its influence was felt, the movement was arrested. In the most enlightened parts of Germany, the seat and home of Protestantism, the 'Franklin rods' early made their appearance. The first lightning conductor set up over a public building in Europe was erected early in 1769 on the steeple of the church of St. Jacob, Hamburg; and so rapid was the spread of them that, at the end of five years from this date, there were estimated to be over seven hundred conductors within a circle of ten miles of the old Hanse town. To this day they are comparatively more numerous in this district than anywhere else in Europe. In contrast with Northern Protestant Germany, the Roman Catholic South refused the 'Franklin rods,' and so did France, although making a hero of Franklin personally. For many years after young Robespierre pleaded the case of lightning conductors before the tribunal of St. Omer, the strongest abhorrence to them was expressed by the priests and their mob following in almost all parts of France, and the active antagonism did not cease till after the outbreak of the great revolution.

It was the same in most countries of southern and central Europe. Even in Geneva, famous for the enlightenment of its citizens, the populace made an attempt to pull down the first lightning conductor. It was erected, in the summer of 1771, by the celebrated naturalist, Professor Horace de Saussure, over his own house, after directions furnished by Dr. Franklin. But notwithstanding that the professor was himself highly respected, his lightning conductor created general abhorrence, and to appease it he found it necessary to issue a public address or 'manifesto,' as he called it, to his fellow-citizens. The address, dated November 21, 1771, was strangely characteristic of the times. 'I hear with regret,' Professor de Saussure declared, 'that the conductor which I have placed over my house to protect it against lightning, as well as to observe, occasionally, the electricity of the clouds, has spread terror among many persons, who seem to fear that by this means I draw upon the heads of others those dangers from which I myself wish to escape. Now, I beg you to believe that I would never have decided upon erecting this apparatus, if I had not been fully persuaded both of its harmlessness and its utility. There is no possibility of its causing damage to my own house, or of doing harm to others. All those who are now labouring under fear would be precisely of the same opinion, if they had entered upon the same inquiries to which I am called in the course of my studies.' After which the professor goes on minutely to describe the 'electric conductor,' which he had been bold enough to place over his house, dwelling upon the fact of its having protected, as he believed, already his own residence from being struck by lightning, and of having been found, likewise, universally efficacious in the same manner in 'the English colonies of North America.' The citizens of Geneva, much given to reasoning, earnestly read and studied the 'manifesto' of Professor de Saussure, and the consequence was, not only that he was spared further attacks and reproaches, but that there arose soon over the churches and houses of the town some hundreds of lightning conductors.

In Italy the progress in the erection of conductors was accompanied by some very curious incidents. The priests here, as in other Roman Catholic countries, actively opposed their introduction, and to do so more effectively, they craftily attached to them a stinging name, calling them 'heretical rods.' As a consequence, the mob fiercely opposed the putting-up of any such accursed pieces of metal, and whenever the attempt was made to fasten them to houses, it met with forcible opposition. However, some of the highly accomplished professors of the universities of Italy, enthusiastic in their reception of Franklin's discovery, proved themselves victorious over both priests and mob. They got the Grand Duke Leopold of Tuscany—subsequently German Emperor, under the title of Leopold I.—a man of high scientific acquirements, to place lightning conductors over his own palace, as well as over all the powder magazines in his dominions. Here the mob and priest rule ceased, and only silent curses could be levelled against the 'heretical rods.' Another still more important step in advance was made by the influence of the Abbé Giuseppe Toaldo, a warm admirer of Franklin, in correspondence with him, and author of various scientific works, among them one on lightning conductors. He had some influence with the ecclesiastical authorities at Siena, in Tuscany, and brought it to bear upon them by getting them to consent to make trial, in a manner so as not to excite public attention, of one of the 'heretical rods,' over the cathedral. This was only permitted on account of the extreme danger in which the edifice stood, having been struck several times by lightning, and greatly damaged. Placed on the summit of the highest of the three hills on which stands the ancient city of Siena, the cathedral was opposed to the dangers brought in the womb of every passing thunderstorm, and they were all the greater as the building, erected by Pisano in the thirteenth century, was deemed to be priceless, being one of the most magnificent structures of the kind in Italy, of red and white marble, filled with the choicest specimens of art, statues, pictures, gold and jewelry. It seemed well worth risking a little heresy to guard such treasures.

Very silently, in the dark of night, the priests of the Siena cathedral, directed by Abbé Toaldo, laid their iron rods along the walls of the building, but inside, planting them deep into the ground, and with the pointed summit only a few feet above the highest point of the steeple, so as to be scarcely perceptible from below by the naked eye. Still the secret of what had been done could not be entirely kept from the multitude. Some of the workmen, engaged in the operation of fixing the iron rods to the inner walls and steeple of the cathedral, whispered about what they had been doing, trembling at the evil consequences of their work, notwithstanding having received full absolution from their employers. Murmurs were now heard everywhere, and there were signs of a popular outbreak, just when one of the many thunderstorms regularly visiting the mountain city crept over it on April 18, 1777. Portentously the black clouds laid themselves thicker and thicker over the high cathedral, till all the people of Siena crept forth from their houses, awaiting in breathless expectation the terrors to come. Then the dark masses discharged their fiery streams; flash followed flash, till one, a long hissing tongue of flame, fell down upon the cathedral steeple, distinctly visible to thousands of beholders. A few minutes after, a ray of sunshine pierced the dark clouds, and to the bewildering astonishment of the masses, the cathedral was standing there absolutely unhurt. As if to exhibit its wonderful power, the gilded point of the lightning conductor stood out brilliantly in the sun, pointing in radiant silence up to heaven. 'Maraviglia, maraviglia!' cried people and priests in chorus. High mass was held forthwith in the wonderfully preserved cathedral, and on the same day the magistrates of Siena went into the town hall and had a record made in the book containing the annals of the city, to make known to all posterity that their noble cathedral had just been preserved from destruction by the astounding influence of an 'heretical rod.' Though not in the least intended to be sarcastic, the irony could not have been more complete.

There was a most remarkable historical concurrence between the gradual introduction of lightning conductors into Europe and that of the art of vaccination. Both the great scientific discoveries had the same end in view for the benefit of mankind, the one teaching the art of drawing the dangerous electric fire of the clouds harmlessly into the earth, and the other that of extracting the poisonous seed of disease from the human body. Both were brought forward with the noblest intentions; and both encountered the most violent opposition from religious fanatics, the same in substance, as interfering with the decrees of Providence, and the ordained wrath of heaven. Both triumphed in the end, and almost exactly at the same time, though the battle of the great medical discovery lasted longer, and was more fiercely fought than that of Franklin's invention. To make

the analogy between the progress of lightning conductors
and of vaccination complete, it so happened that in at least
one conspicuous instance the same man was an important
agent in forwarding the success of both discoveries. The
person in question was Dr. Johan Ingenhousz, a native of
Breda, in the Netherlands, born in 1730. A man of great
natural gifts, he came to England when about thirty years
of age, practising as a physician, and attending specially to
the so-called Suttonian method of inoculation against the
small-pox, then an entirely new branch of medical science.
At the same time he eagerly embarked in electrical experiments, got into correspondence with Benjamin Franklin, and,
having made many friends, was elected a fellow of the Royal
Society in 1769. Recommended to the king, Dr. Ingenhousz
became a favourite at court, owing chiefly to his perfect
knowledge of German, which resulted in his being recommended to a highly profitable as well as distinguished
mission. The famous Imperial lady, the Elizabeth of her age,
Maria Theresa of Austria, had read of the benefits of vaccination, then chiefly known in England, and wishing to confer
them on her own family and friends, she asked King George
the Third to recommend to her some able physician, who
could come to Vienna for the purpose. His Majesty at
once named Dr. Johan Ingenhousz, a recommendation
warmly supported by the President of the Royal Society, Sir
John Pringle, who had taken an affection for the young
Dutch physician on account of his electrical researches,
which had resulted in the invention of a novel apparatus,
subsequently known as the plate electrical machine.

Dr. Ingenhousz set out for Vienna in 1772, was received
with marked honours by the great Empress, and having
done his work, and wishing to visit Italy, received an
autograph letter of Maria Theresa to her son, Grand Duke
Leopold of Tuscany. At the court of this enlightened
prince, Dr. Ingenhousz resided for some time, practising
vaccination, but also engaged in electrical experiments,
which created the greatest interest. It was partly by his

advice that the Grand Duke consented, in the teeth of desperate priestly opposition, to erect one of Franklin's lightning conductors over his own palace, and to set them up likewise for the protection of all the powder magazines in Tuscany. This done, Dr. Ingenhousz went forward to Padua, invited by some of the professors of the university, and by the famous senator of Venice, Angelo Querini, who had a magnificent palace in the neighbourhood of the city. In this palace, bearing the name of Altichiera, the 'English doctor,' as he was called, was made to reside, practising vaccination, the same as at the court of Florence, but following as a favourite occupation the setting-up of 'heretical rods.' Altichiera itself had the first erected in May 1774, and soon after Dr. Ingenhousz had the satisfaction of planting another over the astronomical observatory of the university of Padua, in the presence of an enormous crowd of students who lustily applauded, and of an angry multitude, kept in the background less by persuasion than the strong arms of the young men. As at Siena, so at Padua, the mob became pacified not long after by seeing the lightning fall upon the observatory, much exposed by its situation, and which had often been struck before, without doing the least damage. From Padua, Dr. Ingenhousz went to Venice, in company of his friend and patron, Senator Angelo Querini. Here his efforts to spread the knowledge of lightning conductors, together with vaccination, had the best results. The church of St. Mark and other public buildings were surmounted before long by the awe-striking 'heretical rods,' and on May 9, 1778, the Senate of Venice issued a decree ordering the erection of lightning conductors throughout the republic. It was the first recognition of the value of conductors by any government of Europe, or, indeed, of the world.

CHAPTER V.

METALS AS CONDUCTORS OF ELECTRICITY.

In the history of human inventions and discoveries, the idea of the lightning conductor is almost the sole one which sprang, all but perfect, from one brain, like Minerva, in Greek mythology, from Jupiter's head. Benjamin Franklin discovered the lightning conductor, and, except some important improvements in its manufacture, due to the progress of the metallurgical arts, the conductor remains the same, in essence, as designed by the world-famous citizen of Philadelphia. The reason of this is plain enough. Though one of the most brilliant discoveries in the annals of mankind, the lightning conductor, by itself, is one of the simplest of things. Franklin found by experiments, that the mysterious so-called 'electric fluid' had a tendency to make its way in preference through metals, and so he recommended the laying-down of a metallic line from the clouds to the earth to prevent damage to surrounding objects, such as buildings and the human beings within them. More than this he did not know; and more than this we, to this day, do not know. Of the inner nature, or constitution, of that grand cosmic discharge of electricity to which the name of lightning is given, no scientific explanation can be given. We are utterly ignorant of it, and in all probability ever will be.

But while the general principle laid down by Franklin, that metals will conduct the electric force harmlessly from the clouds to the earth, remains the same, very much has been learnt, in the progress of scientific investigation, as regards the varying conducting capacity of different metals. The first conductors were invariably rods of iron, this metal being preferred by Franklin and his immediate followers as cheap, ready at hand, and answering all purposes in practice. But it was gradually found by experiments that there are other metals through which the electric force will make its way more rapidly than through iron. One of the earliest investigators of this subject was Sir Humphrey Davy, the celebrated inventor of the miner's safety lamp. It was while studying the decomposition of the fixed alkalies by galvanism, and tracing the metallic nature of their bases, to which he gave the names of sodium and potassium, that the great chemist and natural philosopher was brought to enter upon an examination of what may be called the permeability of the different metals by the electric force. The result of his investigations, as stated by him, was that silver stood highest as a conductor of electricity; next to it coming copper; then gold; next, lead; then platinum; then the new metal called palladium—discovered by Wollaston, 1803, in platinum—and lastly, iron. These were the principal metals experimented upon by Sir Humphrey Davy, and the net result of his inquiries was expressed summarily in the fact of copper being more than six times, and silver more than seven times, as good a conductor as iron. Taking copper at 100, Sir Humphrey Davy drew up the following table of the electrical conductivity of the seven metals:—

Silver	109·10
Copper	100·00
Gold	72·70
Lead	69·10
Platinum	18·20
Palladium	16·40
Iron	14·60

The practical result of these experiments was that it came to be recognised that, among the metals, copper might be employed to greater advantage as a lightning conductor than iron: a much lesser substance of it doing the same service of passing a given quantity of electricity from the clouds harmlessly into the earth.

Sir Humphrey Davy was followed in his researches on the conductivity of the different metals by the electric force, by a number of other scientific men. His immediate successor in entering upon this line of observations was a French naturalist of eminence, Antoine C. Becquerel. Perhaps no man after Benjamin Franklin studied the phenomena of electricity with such thorough insight, free from all misleading theoretical delusions, as Becquerel. He was educated at the Polytechnic School of Paris, and in 1810, at the age of twenty-two, entered the army as an officer of engineers, but quitted it five years afterwards with the rank of colonel, to devote himself entirely to scientific pursuits. Geology and mineralogy first engaged his attention, but he soon quitted these studies to devote himself, heart and soul, to the observation of the phenomena of electricity, which fascinated him as much as they had done Benjamin Franklin. The result was the discovery of a great many facts previously unknown, making Becquerel, amongst others, one of the founders of the science of electro-chemistry. The result of his researches concerning the conducting power of the electric force by different metals may be stated as follows:

Copper	100·00
Gold	93·60
Silver	73·50
Zinc	28·55
Platinum	16·40
Iron	15·80
Tin	15·50
Lead	8·30
Mercury	3·45

It will be seen, in comparing this statement with the result of the investigations of Sir Humphrey Davy, that while the latter places silver before copper in conductivity, Becquerel puts copper at the head of the list. Probably, the explanation of this difference in the result of scientific research, by two men equally learned and equally able, may be found in the fact that the conductivity of copper varies greatly according to the purity of the metal. It has been ascertained that absolutely pure copper of the finest kind —such as that existing in the Isle of Cyprus, youngest of mother Britannia's colonial children—has a conducting power of upwards of twenty per cent. more than the ordinary copper of commerce. While thus arriving at different estimates, Sir Humphrey Davy and Becquerel are singularly in agreement in one important respect: they both make the relative electrical conductivity of copper and iron about the same, placing it, the one a little under, and the other a little over 100 to 15. In other words, they both say that the value of copper as a lightning conductor to iron is as twenty to three, or between six and seven times as great.

Among a host of other investigators of the subject there stand forward, besides Sir Humphrey Davy and Antoine Becquerel, two Germans, Professors Lenz and Ohm, and another French savant, Claude Pouillet. In the opinion of many scientific authorities, especially in the United States, the experiments of Professor Lenz regarding the comparative electrical conductivity of different metals were more carefully made than any other, and are therefore deserving of the greatest credit. He had, indeed, ample means and great leisure at his disposal, making his scientific investigations under the patronage of the Grand Duke, afterwards Emperor, Nicholas of Russia, while acting as his private tutor at the university of St. Petersburg. The researches of Professor Lenz as to the comparative power of various metals to conduct the electric force were given in the following results—copper, as before, standing as the centesimal unit:—

Silver	136·25
Copper	100·00
Gold	79·80
Tin	30·84
Brass	29·33
Iron	17·74
Lead	14·62
Platinum	14·16

A comparison of the figures here given with those of Sir

Humphrey Davy and of Becquerel shows that the results obtained by Professor Lenz differ from those of both the other investigators. Like Sir Humphrey Davy, Professor Lenz declared silver to be of greater electric conductivity than copper, but, on the other hand, he assigned lead a very low place, putting it under iron, instead of far above it. It is difficult to explain this wide divergence, even on the utmost allowance of purity, or impurity, of metals. As regards the most important question, from a practical point of view—that of the difference between copper and iron—Professor Lenz, it will be noticed, places iron higher in the scale than both Sir Humphrey Davy and Becquerel. Still, in his estimate also, copper was admitted to have about six times the conductive power of iron.

While, as just stated, the experiments of Professor Lenz on the electric conductivity of metals are held in the highest esteem in America, the same is the case in Germany as regards those of Professor Ohm. The latter is held to be there the highest authority on all subjects connected with the measurement of the electric force. The professor, born at Erlangen, 1787, and for many years teacher of natural history at Munich, where he died in 1854, devoted the utmost patience and an immense amount of time to the definite object of ascertaining the electric conductivity of all the metals, registering the result of his experiments in a special work, the most complete existing on the subject. According to Professor Ohm, the principal metals stand to each other in conductivity as follows:—

Metal	Conductivity
Copper	100·00
Gold	57·40
Silver	35·60
Zinc	33·30
Brass	28·05
Iron	17·40
Platinum	17·10
Tin	16·80
Lead	9·70

Here again is a striking difference with the statements of other investigators. It seems absolutely inexplicable indeed, how it could happen that scientific men of eminence, and admitted authorities on the subject they are treating, came to vary on the electric conductivity of several of the metals. The difference is most astounding as regards silver, the conductivity of which, compared with the per cent. of copper, Professor Lenz places at 136·25, Sir Humphrey Davy at 109·10, Becquerel at 73·50, and Professor Ohm at only 35·60. The only conclusions that can be come to under the circumstances are, that the record of Professor Ohm's results as regards silver is incorrect; or, that the relative degrees of purity of the samples of metal experimented upon by him and the other professors differed very widely. What is of more importance than this question, is the comparative rank of copper and iron. Here, it is satisfactory to find, the results ascertained by Ohm agree very nearly with the conclusions of the other investigators, it being laid down that copper has about six times the conductive power of iron.

The place filled in America by Lenz, and in Germany by Ohm, is generally assigned in France to Professor Claude Pouillet, a savant who devoted, perhaps, more time than any other in his own country to the study of the phenomena of electricity. Born in 1791, Professor Pouillet became, at a comparatively early age, the director of the celebrated scientific institution of Paris known as the 'Conservatoire des arts et métiers,' which led him to enter upon a course of experiments in electricity, and most particularly, at the request of the government, upon investigations as to the best material for lightning conductors. The result of these was published in a lengthened treatise, in which Professor Pouillet set down the electric conductivity of the principal metals, taking copper at a hundred, as follows :—

Gold	103·05
Copper	100·00
Silver	81·26
Brass	from 23·40 to 15·20
Platinum	22·50
Iron	from 18·20 to 15·60
Cast Steel	14·75
Mercury	2·60

It will be seen that Professor Pouillet, differing from other investigators, as they among themselves, regarding the relative conductivity of the precious metals, gold, silver, and platinum, agreed in the main with them as regards the relative proportions of copper and iron. Most painstaking and minute in his experiments, he found moreover that iron, as well as brass—the latter a mixed metal, and as such variable in composition—was not always the same in respect to conductivity, the changes being due to difference in temperature, as well as greater or lesser metallic purity. As set down by him, the variations in iron were between a maximum of 18·20 in regard to 100·00 of copper, and a minimum of 15·60, which gives a mean of 16·90. Taking this mean, the comparative list of the positions held by copper and iron in regard to electrical conductivity, according to the five investigators, may be set forth in the following summary:—

	Copper	Iron
Davy	100·00	14·00
Becquerel	100·00	15·80
Lenz	100·00	17·74
Ohm	100·00	17·40
Pouillet	100·00	16·90

Taking the average of these five statements, it will be found that the relative conductivity of copper to iron stands as 100 to $16\frac{1}{2}$—that is, a little over six to one. The approximate correctness of this figure, being the result of all the investigations by the most eminent men who studied the subject, can therefore admit of no reasonable doubt.

The important researches as to the greatly varying degree in which given quantities of metals will act as conductors of the electric force, were made possible only by the discovery of the singular phenomena of electro-magnetism, due chiefly to the Danish philosopher and naturalist, Hans Christian Oersted. His career, in some respects, was not unlike that of Benjamin Franklin. The son of an apothecary, born in 1777, he set up in the same business, not despising trade, but devoting himself actively to it, as means to an honourable end, that of gaining independence. Fascinated by the study of the phenomena of electricity, Oersted devoted himself to it heart and soul, as Franklin had done; and the result achieved, if not fully as important as the invention of the lightning conductor, was one filling a prominent place in modern scientific discovery. It had been observed, long before Oersted, that there was a close connection between what was known as magnetism and lightning, or rather, to state it more directly, it was known that lightning exercised a strong influence upon the magnetic needle. One of the most notable reports, and one of the first on the subject, came from the captains of two English vessels, sailing in company from London to the West Indies in the year 1675. When near the Bermudas, a stroke of lightning fell upon the mast of one of the vessels, doing considerable damage, and, as the captain believed, swinging his ship round, the men at the helm seeing the compass violently disturbed. He continued steering in what he believed the old direction, but noticed, a few minutes afterwards, that the other vessel, his former companion on the route, and which had not been struck by lightning, was following an opposite course. He had the good sense to approach it, and explanations ensued, the result being the discovery that the lightning had completely reversed the polarity of the magnetic needle, it pointing now south instead of north. The story of this met with much doubt at the outset, but it was amply verified before long by the report of many similar occurrences. It became known, not only that the polarity of the magnetic needle might be reversed by a stroke of lightning, but that the effect of the latter frequently was to magnetise iron and steel. An instance of this kind, on a large scale, occurred at Wakefield, Yorkshire, in the month of June 1731, during a violent thunderstorm. The lightning here entered the warehouse of a merchant who had just packed a case of knives, forks, and other articles of steel and cutlery ware, for despatch to the colonies. The case was placed immediately under the chimney, which the lightning entered, breaking open the box, and scattering over the floor of the room its contents, which, when afterwards examined, were all found to be strongly magnetic. These, and many similar facts, were all clearly established ; yet a considerable time elapsed before important conclusions were drawn therefrom. As in the case of Franklin, so in that of Oersted, it required not merely scientific acumen, but a thoroughly practical mind, to trace, in the one instance, the actual connection between electricity and lightning, and in the other that between magnetism and electricity.

It was in the year 1819 that Hans Oersted, now settled as a lecturer at Copenhagen, announced the result of a series of investigations which laid the foundation for the new science of electro-magnetism. He stated that he had found that if a magnetic needle, free to move like that of a compass, was brought parallel to a wire charged with electricity, it would leave its natural place and take up a new one, dependent on the position of the wire and the needle relative to each other. If the needle, he said, was placed horizontally under the wire, the pole of the needle nearest the negative end of the electric battery would move westward, but, on the other hand, if the needle was placed above the wire, the same pole would move eastward. Again, if the needle was placed on the same horizontal plane as the wire, no motion would be on that plane, but the inclination would be to a vertical movement. Finally, if the wire was laid to the west of the needle, the pole nearest the negative side of the battery would be depressed, but it would be raised if the wire was placed to the east of the needle. From these observations, verified in numerous experiments, Oersted concluded that the magnetic action of the electric force moved in a circular manner around the conducting object, which he expressed in the formula that ' the pole *above* which the negative enters is turned to the west,' and that ' the pole *under* which it enters is turned to the east.' The discoveries of Oersted resulted in the creation of that wonderful production of modern science—the electric telegraph. A minor result, highly important as regards the erection, and still more the maintenance, of lightning conductors, was the construction of galvanometers.

What the microscope is to the student of the inner secrets of animal and vegetable life, the galvanometer is to the investigator of the phenomena of electricity, in their practical applications. Until its invention, there existed no means of practically testing the strength of the electric force, or the 'current,' as it is usually called, and it was not possible, therefore, to ascertain, in any given case, whether lightning conductors, among others, were really efficient or not. Perhaps, had it been only for this purpose, the galvanometer would have waited long in being constructed, but what brought it into existence, and led it to its present perfection, was that greatest of practical uses of electricity, the telegraph. As it arose from small beginnings to gradually more extended employment, embracing ultimately some of the highest interests of civilised mankind, there came the necessity of having instruments for gauging accurately the effects of the mysterious force thus put in harness at the bidding of science. The galvanometer having been devised, the next step, indispensable for its use, was to frame a standard by which electrical energy might be measured, and to invent terms by which the amount of such energy could be expressed. It is well known that in order to be able to measure the dimensions of any material object, standard units are required. In this country the units adopted are: for length, the foot; for weight, the pound; for time, the second; and so on. To express mechanical force or power, the foot pound is the unit employed—that is, the mechanical energy necessary to raise a weight of one pound to a height of one foot. On the Continent, where the units of length and weight are the metre and the gramme, the unit of mechanical energy is the metre gramme. Apart from the fact that the latter units are very generally adopted by all the Continental States, the simplicity of the decimal method of multiplying and sub-multiplying them renders the system of particular usefulness for scientific purposes; and they are therefore very extensively employed even in England in scientific research. Thus experimental results obtained in one country are at once understood, and are directly comparable with results obtained in any other country, without the necessity of reducing the figures to terms of units of other kinds than those in which they are expressed.

Now electrical energy being merely a form of mechanical energy—the one being capable of conversion into the other—it follows that the units of the functions of either of the two powers can be expressed in units of the other; and this being the case, it is manifestly both convenient and desirable that in forming the dimensions of the standard electrical units, they should be constructed in terms of the metre gramme, second units.

The proposition to do this originated with Dr. Weber, and acting upon this proposition a committee of the British Association, comprising nearly all the leading electricians of Great Britain, was formed some years ago, which committee, with almost perfect experimental skill, determined an absolute measure for the values of the several units required for electrical measurement. Taking as the unit quantity of electricity that amount which would be generated by a gramme weight falling through a distance of one metre in one second, the value given to the unit of resistance was such as would allow this unit quantity to flow through it in one second. The means by which the values were experimentally arrived at cannot be described here. It suffices to say, that the unit of resistance being once determined, copies of it, formed of lengths of wire of a platinum-iridium alloy, were issued, from which copies the sets of resistances now so largely employed by electricians were adjusted. Out of compliment to the great German physicist who first proposed the fundamental law which governed the flow of the electrical current, the unit of resistance was called the 'Ohm.' It was a marked progress on the practical application of the electric force to be enabled to measure it, and, as it were, bring it under control.

Without its help the electric telegraph could not have become what it is; nor has it been without notable use in the art of protection against lightning. One of the greatest steps in advance in the application of the lightning conductor, from its discovery to the present day, has been the invention of the galvanometer. Franklin could not, but we can, test our lightning conductors.

Some of the simplest and most practical galvanometers,

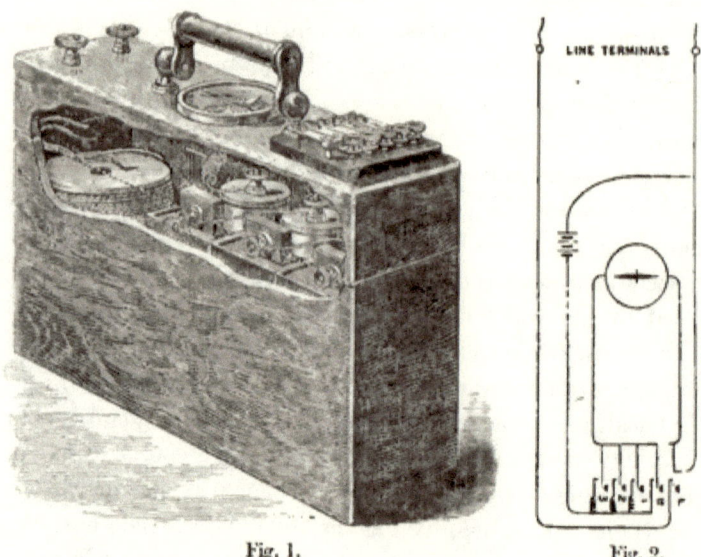

Fig. 1.   Fig. 2.

specially designed for ascertaining the actual efficiency of conductors, have been made in recent years in Germany. The author of this work had constructed for him by Mr. H. Yeates, of Covent Garden, the one, with some improvements, as shown in the subjoined engravings: the first, fig. 1, exhibiting the arrangement of the battery and resistance coils, and the second, fig. 2, giving a diagram of the battery current. The battery consists of three cells, and is a modification of the old manganese cell, in which the carbon and oxide of manganese occupy the outer, and the zinc plate the inner, or porous, cell. By this arrangement, the surface of the negative element is greatly increased, and hence a more constant current is obtained, on account of the battery not polarising so rapidly as in the old form. Another advantage of this arrangement is, that the cells can be almost entirely sealed up, the air-openings being made within the porous cell. In the centre of the lid of the box is placed the galvanometer with a 'tangent' scale; and on the left are two terminals, by the connection of which the conductor can be examined. On the right hand end of the lid are placed five keys, marked respectively, L, B, 1, 2, 3. Under B is one pole of the battery, so that by depressing this key, as will be seen by the connections in the diagram (fig. 2), the battery current is sent through the galvanometer direct. If, however, key No. 1 is depressed, the battery is connected with the galvometer through a known resistance—key No. 2 has a larger resistance, and No. 3. still larger. The fifth key, L, closes the circuit within the limit of the instrument, but on being depressed opens it, and includes the line or conductor placed between the two terminals at the other end, the battery key at the same time being pressed down. By this arrangement it will be seen that the resistance of the line or conductor may be compared with the known resistance connected with any of the keys Nos. 1, 2, 3, or any of these resistances may be included with that of the line, so as to get a convenient deflection of the galvanometer needle. In the case, with the battery, is a bobbin of insulated wire for connecting the instrument with the conductor and earth which is to be tested. The whole arrangement here described and illustrated is exceedingly portable, being in the form of a small carpet bag, and therefore particularly fitted for persons inspecting lightning conductors and making periodical tests, without which it cannot be too widely known there is really no trustworthy security of protection in lightning conductors.

CHAPTER VI.

CHARACTER OF LIGHTNING AND OF THUNDERSTORMS.

It is well remarked by Arago, that although we know nothing about lightning, beyond the well-ascertained fact that it is one of the manifestations of the equally vast and mysterious electric forces pervading the universe, we yet may ascertain a great deal about its mode of action by continued observation, made by many persons and at many places. As yet the wise recommendation of the French astronomer has, unfortunately, not been acted upon to any extent, or in any systematic manner; still, a good many facts and incidents have been gathered which serve to throw a strong light upon the apparently erratic, but in reality normal manner in which, as in obedience to some grand unfathomable cosmic law, the fire of the clouds flashes along its self-made path. That these observations are entirely modern, detracts nothing from their value. With all their famed civilisation, the classical nations of the ancient world never came to look upon lightning and thunderstorms as regular functions of nature, but regarded them with dread and horror. Even the greatest of their natural philosophers found in them means only for encouraging popular superstition. Thus Pliny the Elder, in his celebrated 'Natural History,' recommends, like Arago, notes being taken about thunderstorms, but for quite a different purpose. 'Nothing is more important,' says the celebrated author of the *Historia Naturalis*, than to observe from what region the lightnings proceed, and towards what region they return.

Their return to the eastern quarter is a happy augury. When they come from the east, the prime quarter of the Heavens, and likewise return thither, it is the presage of supreme felicity.' It is reported by travellers that this form of superstition—which has reference, of course, to the zigzag form of many strokes of lightning, apparently in turn advancing and retrograding—still exists in some districts of Southern Italy.

The superstitious awe with which lightning was looked upon not only in ancient times, but in which it is still held by the ignorant at the present day, finds its easy explanation both in the nature of the terrifying phenomenon, and in the fact that even now we can only speculate upon some of the causes of its seemingly capricious actions. There can be no doubt that thunderstorms will visit some districts in preference to others, and that lightning will descend constantly upon some selected spots, and will entirely keep away from others. As regards the latter case, old historians were fond of quoting the grand temple of Solomon at Jerusalem, which was never struck by lightning in the course of a thousand years, although thunderstorms burst unceasingly over the Holy City, creating immense havoc and destruction. In this instance at least, the explanation is simple, although it may not be so in many others. It is stated expressly in the biblical description of the building of the world-famed temple (1 Kings vi. 21, 22) that 'Solomon overlaid the house within with pure gold; and he made a partition by the chains of gold before the oracle; and he overlaid it with gold. And the whole house he overlaid with gold, until he had finished all the house; also the whole altar that was by the oracle he overlaid with gold.' If wise King Solomon had known Franklin's discovery of the protection against lightning given by metallic conductors, he could not have guarded his magnificent edifice better than he did by having 'the whole house overlaid with gold,' as stated in the Bible. But he did even more than this, according to the historian Josephus, who records that the roof of the Temple, constructed in what is now called the Italian style, was ornamented from end to end with sharply pointed and thickly gilded pieces of iron, in lancet form. These points, the historian surmised, were placed there to prevent the birds from settling on the magnificent roof, and soiling it, and it is very possible that this was the original design. Nevertheless, it is certain that King Solomon guarded, although, probably, without intending to do so, his magnificent temple as perfectly against lightning, as could have been accomplished by the best arranged system of conductors. It is not often that many thousands of pounds are spent for protection against lightning, even if intended for great cathedrals and splendid royal palaces; but King Solomon disbursed, by the most trustworthy calculations, no less than thirty-eight millions sterling in covering the temple with one of the best of conductors—including the pointed and gilded lancets along the roof, as perfect 'Franklin rods' as were ever designed by any architect.

If it is easy to account for the old historical marvel of Solomon's temple having stood unharmed amidst the ragings of lightning from tens of thousands of storms, it is more difficult to find the reason why many buildings of another kind should be constantly under attack. A notable case in point, related by the German naturalist G. Ch. Lichtenberg, occurred at the village of Rosenberg, in the province of Carinthia, Austria, belonging to the noble family of Orsini. The village church, although not standing in a very elevated position, was unceasingly struck, in the course of the seventeenth and eighteenth centuries, by lightning, which sometimes battered in the roof, sometimes broke down part of the steeple, and often flew in at the window on one side and out on the other. Very possibly, there were large pieces of metal on the wall, or in the roof; or, if not, there may have been masses of water near, underground, sufficient to account for the manifestations of the electric force. However, popular opinion, utterly ignorant as to such causes, ascribed the whole to the doings of evil spirits, and endless attempts were made to exorcise them by prayers, fastings, and sprinkling of holy water. But it was all unavailing. The lightning came again and again, and in the summer of 1730, a flash from the clouds, more violent than any preceding one, demolished the entire steeple. The Orsini family, suspected by many of the lower people of being the secret originators of all the mischief, in league with the evil spirits, erected another steeple, handsomer and far more solid than the one destroyed, to show their pious intentions. But the lightning visited it as before, on the average five or six times a year, doing so much damage that the whole church had to be taken down in 1778, being found in ruins. Happily, in the meanwhile the report had gone as far as the little village in Carinthia that something had been discovered for guarding all buildings, including spirit-haunted churches, against damage by lightning. Once more, the proprietors of the village built a new church on the old ground, but this time, by the advice of an Italian architect, they placed upon it one of the 'heretical rods,' made famous for having done good service in protecting the cathedral of Siena. Needless to say that it did again similar service.

It is very probable that, besides the two causes just referred to which divert the path of lightning, there are many others of influence, at present entirely unknown. Numerous cases are reported where the electric discharge from the clouds touched precisely, and with singular accuracy, as if directed by a superior intelligence, the same spot, without the slightest reason being discoverable for such an action, after the most minute investigation by competent persons. Thus, on June 29, 1763, a violent thunderstorm broke over the village of Antrasme, near Laval, in France, the residence of a distinguished investigator of electrical phenomena, Count de Labour-Landry, a friend and correspondent of the Abbé Nollet as well as of Benjamin Franklin. The lightning struck, as carefully ascertained by the Count, first the steeple of the church, then sprang to one of the lower walls, where it fused and blackened the gilding of picture frames and some other metallic ornaments, melting also some pewter flasks used for sacramental purposes, and finally opened two deep holes, as regular as if they had been drilled with an auger, in a wooden table, placed within a recess of soft stone. All these damages were duly repaired; but, to the boundless surprise of the witnesses, the lightning struck the church almost exactly a year after, on June 20, 1764, entering the church by the same way as before, fusing and blackening the same gildings, melting the same flasks, and, in the end, driving out the very plugs of the wooden table inserted to fill the holes bored by the previous stroke of lightning. The account of the whole might seem almost incredible, were it not attested by independent eye-witnesses. Arago, with absolute faith in their testimony, remarks thereon that 'those who will take the trouble of reflecting upon the thousands of combinations which might have caused the path of the lightning to have been different in the two cases, will, I imagine, have no hesitation in viewing, with me, the perfect identity of effect as demonstrating the truth of a proposition I put forward,' namely, that 'lightning, in its rapid march, is influenced by causes, or actions, dependent on the terrestrial bodies near which it explodes.' In other words, lightning, like many other phenomena of earth, air, and water, is influenced by unknown causes. Hamlet says very much the same, when exclaiming:

> There are more things in heaven and earth, Horatio,
> Than are dreamt of in your philosophy.

After all possible explanations, Arago could get no further than Horatio, who thought that there was much in the universe which was 'wondrous strange.'

It does not seem impossible that some of the extraordinary effects of lightning, either in striking repeatedly certain objects, or in seldom or never touching others, may be explained on meteorological grounds. The height, as well as thickness, of the clouds charged with the electric fire, must naturally greatly influence the direction of the latter, and though both elements vary enormously, in different countries and at different seasons, it is likely enough that the variation is comparatively trifling under given conditions, as, for example, in a district where there are prevailing winds, and where the configuration of the earth powerfully acts upon the drift and movement of the aerial masses, and the atmospheric conditions in general. As to the height of the clouds charged with lightning, there appears scarcely any limit, as it has been found that they at times rise above the summit of the most elevated mountain ridges on the face of the globe. The great naturalist and traveller, Alexander von Humboldt, measured the height of a storm cloud, discharging lightning, near the mountain of Toluca, in Mexico, and found that it was no less than 4,620 metres, or 15,153 English feet, above the level of the sea. The height of another, ascertained by Professor de Saussure, of Geneva, when ascending Mont Blanc, was 4,810 metres, or 15,776 English feet, or almost exactly three miles. Probably, these are exceptional cases, as even in mountainous districts the heavy moist and electricity-laden clouds seldom rise to such extraordinary heights; still, even such as they are they do not represent the extreme limit of elevation. A member of the French Academy of Sciences, M. De Lisle, records having measured, by trigonometrical observations, the vertical height of clouds in a thunderstorm, with strong flashes of lightning, which broke over Paris, and found it to be 8,080 metres, or 26,502 English feet. Consequently, this cloud-mass, charged with electricity, stood far above the summit of the highest mountain peak in the world.

If some of the lightning-clouds tower at a gigantic elevation over the earth's surface, there are others that lie almost flat on it. There are some remarkable observations on this kind in existence, made by German meteorologists. Two of these deserve particular notice. On August 27, a heavy storm burst over the town of Admont, on the river Ens, in Styria, and the lightning, falling upon the lower part of the great convent of the Benedictines, and passing through the wall, killed two young priests near the altar, while reading vespers. The convent lies, like the town of Admont, in a valley, and above it, some three hundred feet higher, stands a castle, in which resided at the time a German professor, specially interested in the phenomena of thunderstorms. He watched assiduously the coming storm, and saw the lightning fall upon the great convent, noticing all the while that the gilded cross placed on the belfry of the edifice, about 115 feet from the ground, remained standing out clear above the electric cloud, which appeared to come close to the earth's surface. He noticed further that above this cloud, enveloping the ground portion of the Benedictine convent, there hovered another, more than two thousand feet higher, and at intervals he could see streaks of lightning fly from between the two, not however from the more elevated to the lower one, but in a contrary direction. It was evident that the two clouds must have been charged by electric forces of different 'degree,' or 'potential;' it may be, one of a 'negative' and the other of a 'positive' kind, or, as Benjamin Franklin termed them, 'plus' and 'minus;' although, as long as the forces differ in 'degree' or 'potential,' it is not essential that they be of opposite kinds. As the marvellously sagacious discoverer of the lightning conductor surmised, the wondrous force is really unitarian — that is, throughout the same, the term 'kind' really only indicating on which side of an assumed zero (the potential of the earth) the observations or measurements are made.

Another notable instance of low-lying storm-clouds, and which furnished the rare opportunity of measuring the thickness of one of them, occurred at the city of Gratz, Austria, on June 15, 1826. The city is built along the side of a hill, the highest point of which, called the Schlossberg, has on its summit a castle, now in ruins, but at the time garrisoned by troops, and furnished with a small observatory. When the storm in question broke over the city, several scientific men on the Schlossberg took notes of the movement and direction of the great cloud emitting its electric discharges. This they could easily do, as they themselves, on their altitude, were standing in sunshine, under a perfectly blue sky, the dark cloud-wave rolling deep under their feet, indicating its path and size by streams of fire, following each other in rapid succession. Exclusive of short flashes, vanishing in the air as soon as seen, there fell nine great strokes of lightning upon buildings in the city, in the course of about three quarters of an hour, five of them causing conflagrations and killing a number of people. The storm over, the observers compared their measurements, and it was then found that the height of the storm cloud had never been above the clock-tower of the Johanneum, an edifice connected with the university, and containing a library and museum, while the lowest part of it had gone down the sloping ground of the city no further than 120 feet under the summit of the clock-tower. This, then, was the exact thickness of the storm-cloud which had caused so much destruction. It was noticed on this occasion, as had been done often before, that the discharges of lightning fell all upon buildings standing on moist ground, near the river Mur, a mountain stream coming from the Noric Alps, and dividing the city into two parts. There can be no doubt, from thousands of observations made, that it is one of the characteristics of the electric force to seek its way towards water—to be, as it were, dissolved by it, or, as perhaps it might be said more truly, to be equalised by it. A very remarkable electrical phenomenon, and one which is often attended with fatal results to men and animals, is what is known as the 'return stroke' of a lightning discharge. This is always less violent than the direct stroke, but is nevertheless very powerful. It is caused by the inductive action which a thunder-cloud exerts on bodies placed within the sphere of its activity, and disastrous effects often take place upon objects, upon men and animals on the earth under the cloud, although perhaps miles away from the point where the discharge takes place. These bodies are, like the ground, charged with the opposite electricity to that of the cloud;

but when the latter is discharged by the recombination of its electricity with that of the ground, the induction ceases, and all the bodies charged by induction return to a neutral condition. The suddenness of this return constitutes the dangerous 'return stroke.'

Lord Mahon was the first to demonstrate by experiment its mode of action; as shown in the following illustration.

A B C is the electrified cloud, the two ends of which come near the earth. The lightning discharge occurs at c. A man at F is killed by the return stroke, while those at D, nearer to the place of discharge, but further from the cloud, receive no injury. It may be mentioned that it was the action of the return shock upon the limbs of a dead frog in Galvani's laboratory that led to the Professor's experiments on animal electricity, and further to the discovery by Volta of that form of electrical action which bears his name.

The subject of the origin of atmospheric electricity has at all times been a favourite source of speculation with scientific investigators, and given rise to numerous hypotheses. The eminent Swiss savant, Professor de Saussure, already referred to, held that all atmospheric electricity was due to the evaporation of the waters of the globe through the effect of the sun. To prove this, he made a great number of experiments, showing that whenever water, whether pure or containing more or less salt, whether acid or alkaline, is projected upon a metal crucible heated to redness, the evaporation that takes place immediately is accompanied by strong liberation of electricity. The fact is undisputed by scientific men, but not so the conclusion. Another eminent savant, no less distinguished than De Saussure, Professor De la Rive, in taking up the experiments of the former, succeeded in showing that the production of atmospheric electricity by throwing water upon heated metal was not the simple effect of evaporation, but due to chemical causes.

Of the numberless attempts made to elucidate the phenomena of electricity, in connection with the formation of thunderstorms, none seem more worthy of regard, and of thoughtful consideration, than those of Jean Athanase Peltier, a French savant, little known to the general world. Born in 1785, he occupied his whole life, until his death in 1845, with the study of meteorology and electricity, making, among others, the important discovery that a current flowing through a circuit composed of two metals joined together heats or cools the junction according to the direction of the current. From all the experiments upon the phenomena of electricity, to which he devoted his life, Peltier drew the conclusion that the earth itself, and more particularly the fiery liquid mass forming the inner bulk of it, over which the solid crust and the ocean lie, but both thinner in comparison than the skin of an apple, form one immense reservoir of electricity. As light comes from the sun, generated, as we believe, by heat, so the electric force, he held, comes from the interior of the globe, likewise generated by heat. The atmosphere surrounding the globe, Peltier asserted, produced no electricity whatever, nor held it, except temporarily. But he thought it possible that it might exist, engendered by other flaming masses than those of the earth's interior, in the interminable planetary spaces, which no astronomer can measure, and of which imagination itself, in its loftiest flights, can form no more conception than the finite ever can of the infinite. On the whole, Peltier's explanation, such as it is, may fairly be accepted, in the present state of the scientific investigation, as one of the best that can be given. For the rest, men must content themselves to study the phenomena of electricity, and to regard it simply as one of the great, if mysterious, forces of nature.

CHAPTER VII.

INQUIRIES INTO LIGHTNING PROTECTION.

FROM our present ignorance of the actual nature of electricity, admitted alike by all scientific men, it has often been argued that no claim can be set up for a perfect protection against the effects of the electric force called lightning, since we do not know 'whence it comes, nor whither it goes.' That this argument is entirely fallacious, may be easily shown. The human mind does not understand, any more than it does electricity, the great forces called centripetal and centrifugal, which keep millions of suns and of planets in their path through the boundless universe; yet there is no educated man who doubts that astronomers are able to calculate, with the greatest mathematical precision, the time when two particular stars will come near each other, when the moon will obscure Orion, and Venus make her transit across the sun. Again, no explanation can be given of the actual nature, of the Why and the Wherefore, of the force called gravity, simply in its operation on our globe. Still men can calculate, with the greatest nicety, the result of any given weight, falling, from any given height, on the surface of the earth or below it.

François Arago, reasoning on the disputed efficiency of lightning conductors, puts another indisputably practical case. 'If,' says he, 'we take the dimensions to be given to conductors from experience, and if those which we adopt have been found to resist the strongest lightning recorded for over a century, what more can reasonably be asked for?'

When the engineer decides on the height and width of the arches of a bridge, the vault of an aqueduct, the section of a drain, and similar constructions, what does he concern himself with? He examines all the facts and records on the matter as extensively as he can, and, in making his plan, keeps somewhat beyond the dimensions dictated by the greatest floods and the heaviest rains which have ever been observed. He thus goes as far back in his research as the evidence within his reach will enable him to do, but without confusing himself either with searching for the hidden causes of floods and rains, or with investigating the character of the physical revolutions, or the cataclysms which occurred in prehistoric times, and of which geologists only have been able to discover the traces and estimate the magnitude. So with the engineer. Greater precaution or foresight than his cannot be demanded from the constructor of lightning conductors, nor is any needed.'

It may be laid down as an absolute fact, that a well-made lightning conductor, properly placed, and kept in an efficient state, can never, under any circumstances, fail in its action. Undoubtedly it has happened that buildings to which conductors were attached have, in many instances—of which some will be enumerated in another chapter—been struck by lightning, and even damaged; but these cases, so far from going against the truth that good lightning conductors are infallible, only serve to prove it. A close investigation of all known instances where the electric force has struck buildings, nominally protected against lightning, shows most conclusively that the conductors placed on them were either inefficient, in some way or the other, or did not lead properly into moist ground—that is, had not the all-indispensable ' earth connection.' There is no case on record in which a really efficient lightning conductor, properly placed, and with its terminal in technically so-called ' good earth,' did not do its duty; and without being dogmatic on the subject, it may well be asserted can no more fail to give protection than an efficient drain-pipe can fail to carry off the water upon the roof. Although the electric force is neither a 'current' nor a 'fluid,' often as it is so described, still the analogy holds good so far as the one here given between the drain-pipe and the conductor. And the reason is clear enough. The water, in running down a hollow tube, obeys simply the law of gravity, but no less immutable than this is that which governs the movement of the electric force. As the water has no choice but to follow the channel made for it, under the guidance of experience and mathematical calculation, so has the emanation of the electric energy no option but to pursue the path which scientific investigation has shown it always to take. Men may speak of 'erratic' lightning; but it is certain that the course of the electric force is as subject to cosmic laws and as immutable as that of the stars.

Most of the experiments and investigations for ascertaining the best form of lightning conductors, and their application to buildings so as to be invariably efficient, have been carried on by private activity; still, the subject has also, at various times, undergone the examination of official authorities, as well as of learned societies. Little has been done in this respect in England, but very much in France, where, ever since the publication of Franklin's great discovery, the question of protection against lightning has uniformly interested the public, as well as the learned world, leading to the production of more treatises on the subject than in any other country, except perhaps Germany, the world's centre of book-making. One of the most important of the French works here referred to, and which may be regarded as the standard work on lightning conductors, is a semi-official publication, entitled 'Instruction sur les paratonnerres,' issued in new editions from time to time, and widely dispersed, not only in France, but all over Europe and America. It consists of several reports about lightning conductors made, from 1823 to 1867, by committees comprising some of the most distinguished men of science at the time, to the 'Académie des Sciences' of Paris. The earliest of these reports originated from an application of the French

Government to the 'Académie.' In the year 1822, there happened to be in France, and over the greater part of Continental Europe, an extraordinary number of violent thunder-storms, accompanied by earthquakes and simultaneous eruptions of Mount Vesuvius, the latter on a scale not witnessed for centuries. In France, the almost continuous thunderstorms caused great alarm among the population; and the priests in many places held processions in and around the churches, with special prayer-meetings, to 'appease the wrath of heaven.' In consequence of all this excitement, the Minister of the Interior, deeming that something also ought to be done besides the walking in procession to stay the fatal effect of lightning, ordered that all the public buildings in France should be protected immediately by conductors, made on the most perfect model and placed in the best manner. To get pre-eminent advice as to the efficiency of lightning conductors, the Minister applied officially to the 'Académie des Sciences,' which learned body thereupon nominated a committee consisting of six of the most celebrated investigators of the phenomena of electricity — MM. Poisson, Lefèvre-Gineau, Girard, Dulong, Fresnel, and Gay-Lussac. The committee held many sittings, collecting a vast amount of evidence on the subject, and on April 23, 1823, presented through M. Gay-Lussac its report to the 'Académie des Sciences,' which was adopted and ordered to be printed, being declared a highly important document. The French Government took the same view as the 'Académie des Sciences,' and not only acted upon the recommendations of the report, but issued it to all public functionaries, to the clergy, and others, with directions to make it generally known. In this way hundreds of thousands of copies of the 'Instruction sur les paratonnerres' found their way all over France, and from thence in translations all over Europe, as the best existing guide for the erection of lightning conductors.

The information thus spread by the French Government gave rise to important results. It caused the setting-up of lightning conductors throughout the country, on private as well as public buildings, and it likewise led to an improved construction of them, in as far as the 'Instruction' recommended the rods to be made of stout pieces of metal, well fastened to each other, and, above all, led into the ground deep enough to reach moist earth or water. If this was well enough, and useful enough, to meet with general acceptation, there were some points in the advice of the learned men of the 'Académie' that gave rise to much criticism, as being more founded upon theory than practical experience. In the first place, they laid it down as a hard-and-fast rule that the upper rod of a lightning conductor—that projecting over the roof—'will be an efficient protective against lightning within the circular area of a radius double that of its height,'[1] and the acquiescence in this supposed absolute formula had for one of its results the erection of monstrously huge rods, made to tower high above buildings, so as to increase the field of protection to the largest possible extent. Another and worse fault was committed by the authors of the 'Instruction' in not saying anything about the necessity of regularly inspecting the actual condition of lightning conductors, and testing them in respect to their efficiency. While giving minute advice as to the mode of construction and the general design of conductors, the contents of the 'Instruction' were such that, on the whole, its readers would take it for granted that it was only necessary to properly join the strips of metals and bring them down into the ground, after which, thenceforth and for ever, the protection against lightning would be complete. This grave omission, together with the erroneous dogma as to absolute rule of protection within an area prescribed by the height of the 'tige,' or upper part of the rod, had the inevitable result of causing disasters, and before the 'Instruction' had been issued many years, there came report after report to the

[1] The original, long taken as a scientific dogma, runs: 'Une tige de paratonnerre protège efficacement contre la foudre autour d'elle un espace circulaire d'un rayon double de sa hauteur.'

Government that well-constructed lightning conductors had failed to do their duty. For a length of time these reports were either not believed in, or the failure ascribed to partial non-compliance with the strict rules laid down by the 'Académie des Sciences.' However, in the end, when thirty years had passed, the instances of buildings with conductors being struck became so numerous, that it was impossible to ignore them any longer, and, flying once more for advice to the savans of the 'Académie des Sciences,' the French Government desired them to investigate anew the question as to the best means of protecting buildings against lightning. Complying with the behest, the learned body nominated again a committee of six, the names of those selected comprising the most eminent men who had made electricity and its phenomena their study. They were MM. Becquerel, Babinet, Duhamel, Despretz, Cagnard de Latour, and Pouillet.

The 'Instruction' of the new committee, drawn up by Professor Pouillet, was read before the 'Académie des Sciences' on December 18, 1854, and having been unanimously approved, was, like the former one, taken up by the Government and extensively circulated. The report began by modestly excusing the short-coming of its predecessor. 'For the last thirty years,' Professor Pouillet remarked, with no fear of being gainsaid, 'the science of electricity has made great progress—in 1823 the discovery of electro-magnetism had only just been made, and none could foresee the immense results that would spring from its revelations.' Based upon these grounds, the new 'Instruction' entirely reversed many of the conclusions of the old one. First of all, it declared inadmissible the theory of a fixed area of protection, to be calculated by the length of the upper rod. 'Such a rule,' Professor Pouillet justly remarked, 'cannot be laid down with any pretence to accuracy, since the extent of the area of protection is dependent from a mass of circumstances—such as, among others, the shape of the building and the materials entering into its construction. It is clear, for example, that the radius within which the conductor gives protection cannot be so great for an edifice the roof or upper part of which contains large quantities of metals, as for one which has nothing but bricks, wood, or tiles.' Professor Pouillet then proceeded to give detailed instructions in respect to the design and mode of manufacturing lightning conductors. He insisted that the rods should be of greater capacity than those recommended by Gay-Lussac in the report of 1823, and that there should be as few joints as possible from the point to the earth. He considered it of the greatest importance that all the joints should be carefully tin soldered, otherwise the metallic continuity of the conductor could not be assured. He also advised that the top of the air-terminal should not taper to so fine a point as formerly, but be rather blunt. A lightning conductor, said Professor Pouillet, is destined to act in two ways. In the first place, it offers a peaceful means of communication between the earth and the clouds, and by virtue of the power of points the terrestrial electricity is led gently up into the sky to combine with its opposite. In the second it acts as a path by which a disruptive discharge may find its way to the earth freely. In the latter case he considered there was a risk of a sharply tapered point becoming fused, and recommended that the angle of the cone at the top of the air-terminal should be enlarged. He also advised that the point should be made of red copper instead of platinum, and based his argument on the fact of copper being a better conductor of electricity than platinum, and considerably cheaper. A copper point, remarks M. Pouillet, subjected to a heavy stroke of lightning, would be much less heated than a platinum point, and would scarcely in any case be fused. While in the report of 1823, iron ropes were recommended almost exclusively as the best material for conductors for ships, the 'Instruction' of 1854 declared strongly in favour of copper as the far superior metal for the purpose. 'Copper,' affirmed Professor Pouillet, 'is superior to iron as well as to brass for the purpose of lightning conductors, it having the advantage not only of being less influenced by atmospheric agencies, but the still more important one of allowing a freer passage to the electric force of over three to one. Copper should therefore be exclusively used in the construction of lightning conductor cables for the protection of ships.

The inquiries into lightning protection instituted by the 'Académie des Sciences,' and resulting in two reports, the second valuable in the highest degree, had the good effect, not only of drawing public attention to the necessity of providing such safeguards, but of bringing the whole matter under due scientific control. Henceforth the ground was cut away under 'lightning-rod men,' perambulating towns and villages, and offering their trumpery ware—mostly bits of wire tied together, with perhaps a lacquered piece of wood on the top—to credulous persons, as a substitute for good conductors. The French Government set a laudable example in appealing for the future always to scientific aid. A few months after the publication of the 'Instruction sur les paratonnerres,' drawn up by Professor Pouillet, a decision was come to for protecting the new wings of the Louvre, at Paris, with the most perfect lightning conductor that could be made, and thereupon appeal for counsel was once more made to the 'Académie des Sciences.' The case was one of special interest. The palace of the Louvre, with its inestimable treasures of art, had been the first public building in France provided with a lightning conductor. It was due to the initiative of an enthusiastic admirer of Benjamin Franklin, David Le Roy, that this was accomplished, he having excited the public feeling as to the dangers from lightning to which the Louvre was exposed to such a degree as to compel the Government, in 1782, to carry out his plans, under his own superintendence. The conductors erected by Le Roy had stood the test of experience from 1782 until the year 1854, many a thunderstorm having passed over the extensive buildings of the Louvre without causing the least damage. But, in the last month of 1854, one more lightning cloud swept along the banks of the river Seine, and the electric fire, falling on one of the chimneys of the palace, knocked off a few bricks. The damage was very trifling, but the alarm nevertheless was great, and very naturally so. If there was one building in France, it was said, which ought to be beyond the risk of being struck by lightning, it was the Louvre, and, if this could not be accomplished, the art of constructing protective conductors was altogether vain and ineffectual. It was under these circumstances, incited by the public outcry, that the Government hastened to submit the new case to the 'Académie des Sciences.'

Once more the 'Académie' nominated a committee on lightning conductors, composed of the same members who had signed the 'Instruction' of 1854, and drawn up by Professor Pouillet. He again drew up the report, which was adopted by the 'Académie' on February 19, 1855, and contained some notable additions to the directions previously given. They related, as was desired, in the first instance to the Louvre alone, but were made applicable to all large public buildings. For their efficient protection, the professor insisted, two things should be kept in view above all others— namely, first, that the point, always of copper, should be of greater thickness; and, secondly, that it should have a never-failing connection with either water or very moist earth. To ensure the latter, it was recommended, as had been done before, that the underground part of the conductor should be divided.

The necessity for such a division, and for forming at least two subterranean arms—the first of it, described as 'the principal branch,' going very deep into ground, into perennial water, and the second, 'the secondary branch,' running nearer the surface—was explained by Professor Pouillet very clearly in this last report. 'After a long continuance of dry weather,' he observed, 'it often happens that the lightning-bearing clouds exert their influence only in a very feeble manner on a dry soil, which is a bad conductor; the whole energy of their action is reserved for the mass of water which by percolation has formed below it. It is here that the dispersion of the electric force (*la décomposition électrique*) takes place; it will follow the principal branch of the conductor underground, and leave the secondary branch untouched. The case is entirely different when, instead of dry weather, there have been heavy rains, moistening the earth thoroughly, up to the surface. It is the latter now that is the best, because the nearest, conductor of the electric force, which will not go to the more permanent sheet of water, lying more or less deep in the ground, if there is moisture above it. Under these circumstances, it is indispensable that there should be a direct connection between the surface soil and the lightning conductor, and this is what is accomplished by the secondary branch. It is a power in aid of the principal branch, and one often of the highest importance.' The suggestion here made was one so evidently good, that it was at once accepted by the French Government, and the Louvre not only, but other public buildings, received lightning conductors ending in two subterranean branches, as proposed by Professor Pouillet.

The report on the protection of the Louvre Palace did not contain the last inquiry of the 'Académie des Sciences' on the subject of lightning conductors. Twelve years after it had been issued, the Government of France once again called upon that learned body for advice as to the best mode of protecting powder magazines. Several cases had happened—among others at Rocroy, on the borders of the forest of Ardennes—of such buildings being struck, notwithstanding that they had conductors placed upon them, and the Government, naturally alarmed, made inquiry as to whether nothing could be done to ensure protection against lightning, infallible under all atmospheric conditions and every possible emergency, to these dangerous stores. The demand was made in a letter of the Minister of War, Marshal Vaillant, dated October 27, 1866, pressing the 'Académie' to give another 'Instruction,' without delay, the Government being 'in fear that some of the powder magazines are not as completely protected from lightning as could be wished.' Thereupon the 'Académie des Sciences' nominated another commission, this time of eight members, including the Minister of War himself—not complimentary, but as being an author, and with a warm interest in electrical science; and, besides him, MM. Becquerel Sen., Babinet, Duhamel, Fizeau, Edmond Becquerel, Regnault, and Professor Pouillet. The list represented a galaxy of names unsurpassed in the investigation of such a subject as lightning conductors, looked upon in most countries of Europe, at least in recent years, as rather plebeian, to be left to builders and lightning rod men. Many sittings were held by the committee, all fully attended, so that, although the Minister had desired to get the new report '*le plus promptement possible*,' it was not till nearly three months after the receipt of his message that it was completed, Professor Pouillet again being the author. It was a most remarkable paper, this one, read before and approved of by the 'Académie des Sciences' on January 14, 1867.

Before entering upon the subject of the protection of powder magazines against lightning, the new 'Instruction' signed by Professor Pouillet and his colleagues laid down a few so-called '*propositions générales*'—that is, either hints, suggestions, or statements, the French word '*proposition*' being most serviceably vague for use—on the subject of lightning and of thunderstorms. The first thesis affirmed that 'clouds which carry lightning with them are but ordinary clouds (*ne sont autre chose que des nuages ordinaires*) charged with a large quantity of electricity.' The second thesis boldly defined the nature of lightning. 'The fire which flashes from the skies is an immense electric spark, passing either from one cloud to another, or from a cloud to the earth; it is caused by a tendency for the restoration of the electric equilibrium (*la recomposition des électricités contraires*).' It was laid down in the third '*proposition*' that, when lightning falls from a cloud upon the earth, it is but an effort of the electric force to return to its grand reservoir. That it is similar to water, which, having risen in the form of vapour from the earth-surrounding ocean high up into the air, then falls down as rain upon hills and plains, and finally runs down again in rivers to the ocean, Professor Pouillet did not say in so many words; but there were vague hints to that effect in the new 'Instruction.' Its practical recommendation, offspring of the theories thus enunciated, was that the best protection against lightning would be afforded by the most substantial metal rods, made of iron, surrounding a building on all sides, and passing deep into the ground. The new declaration of the 'Académie des Sciences,' though merely a repetition of former reports, was not without important consequences. First in France, and then in other countries, the conviction became general among scientific men, and others well informed on the subject, that well-designed conductors, if properly made and kept in good order, form an absolute, unconditional, and infallible protection against lightning.

Professor Pouillet also laid it down that lightning conductors, to be efficient, must be regularly inspected, he, with his colleagues on the committee, having come to the conclusion that such examination should take place at least once every year. So much stress was laid upon the importance of an annual inspection, that a strong recommendation was made to the Government to have a *procès-verbal*, or special report, drawn up on each occasion in the case of all public buildings, so that it might be known by the central authorities whether the examination had taken place at the specified time, and what had been the declaration of the examiners. The advice was judiciously followed, with the result that at this moment the public buildings of France have the most complete protection against lightning—greatly in contrast with the public buildings in England.

CHAPTER VIII.

SIR WILLIAM SNOW HARRIS.

In singular contrast with what took place in France, the importance of lightning conductors never created any but the most languid interest in England. Neither the Government, nor any of the scientific bodies of the country, at any time occupied themselves seriously with the question as to how public and private buildings might be best protected against the dangers of thunderstorms; and from the time, a century ago, when the Royal Society half patronised and half spurned the merits of Franklin's discovery, to this day, the battle of science against ignorance in the matter had to be fought by individuals. With one exception, that of Sir William Snow Harris, it proved no profitable battle to any man; and in his case even, it was only so by accident. Born at Plymouth, in 1792, and educated for the medical profession, he early turned his attention to the subject of electricity and lightning conductors, and more particularly to the use of them in the Royal Navy. Owing to his early surroundings, leading to connection with naval officers, he learnt that the damages caused by lightning to ships of war were very numerous, and most expensive to repair; and having got once hold of these facts, he gave them to the public in the 'Nautical Magazine,' but chiefly in pamphlet form, insisting upon the simple remedy of lightning conductors. As usual, the Government lent a deaf ear to the proposal as long as it was possible, and it was only when at length, in 1839, the outcry upon the subject became overwhelming, that a naval commission was appointed 'to investigate the best method of applying lightning conductors to Her Majesty ships.' The commission drew up an immense report, filling eighty folio pages of a blue-book, the kernel of which was that, though such protectors in thunderstorms were rather new-fangled things, they might be tried without special harm coming to anybody. Thereupon most of the vessels received lightning conductors, made after designs by Mr. Snow Harris. The indefatigable advocate of conductors had his reward. He was knighted in 1847; he had, at various times, considerable grants from the Government; and he had the final satisfaction of being allowed to design lightning conductors for the new Houses of Parliament. The latter remain the most enduring monument of the only man in this country who ever succeeded in drawing the attention of the public and the Government to the grave subject of lightning conductors. He could not have done so, at least not in the line he took up, had he lived half a century later. With the gradual disappearance of the old wooden ships disappeared also the necessity of lightning conductors for men-of-war. An iron-built vessel, metal-rigged, is a conductor by itself, while as to armour-clad ships of latest design, they are more absolutely protected against lightning even than the famous gilded temple of Solomon at Jerusalem.

In the story of the progress of lightning protection in England, the career of William Snow Harris forms a chapter of no little interest, as showing both the inertness of the administration, as well as of the public, in the most important matters, and the good effects that may result from the persevering energy of a single man. When Mr. Snow Harris began his agitation for lightning conductors, about the year 1820, the ships of the Royal Navy were virtually without them, although they had something supposed to stand in their place. Just sixty years before, in 1762, Dr. William Watson, the indefatigable advocate of Franklin's discovery, had strongly recommended to Lord Anson, first Lord of the

Admiralty, that all men-of-war should have lightning conductors; and his urgent zeal, backed by influential friends, effected that his advice was listened to. Being requested to send in the best design for a ship's conductor, Dr. Watson did so with alacrity, but, unfortunately, with little wisdom. Knowing little or nothing of ships and their management at sea, the learned member of the Royal Society advised that the lightning conductors for the navy should be constructed of strips of copper rod, one-fourth of an inch in diameter, hooked together every few feet by links, and the whole attached, for more security, to a hempen line, to be hung on to a metal spike on the top of the mast, and from thence to fall down into the sea. In theory, it was not a bad design, but it utterly failed in practice. Evidently, Dr. Watson had never been on board of a large ship in a gale, for had he been, he might have known that it would be next to impossible to keep his chain in its place, exposed as it was to the operation of violent mechanical forces, not to speak of possible bad treatment from indignant sailors, with whose movement in the rigging it interfered. It was a natural consequence of Dr. Watson's ignorance, that his conductors entirely failed. In most cases the commanders of men-of-war, supplied with the copper-hempen chains, quietly stowed them away in some corner of the ship, with orders to take them out when needed, and it often happened that this was done only after the ship had been struck by lightning. Year after year there came reports of such casualties; and at last they got so numerous as really to attract the attention of the naval authorities. Still, nothing was done until William Snow Harris took up the matter. Sitting in his little cottage at Plymouth, overlooking the sea, the happy thought struck the young medical man, waiting for patients who did not come, that here might be found a profitable as well as useful opening for his activity. He possessed, happily, a few naval friends, ready with counsel and assistance, and so he went to action, fighting for lightning conductors.

The battle, resulting as it did in ultimate victory, was a

long one, nevertheless. For many years, all his efforts to induce the British Government to adopt a system of efficient lightning conductors for the Royal Navy remained entirely fruitless ; and it was only after he had gained the sympathy of the press, and, through it, of the public, by publishing long lists of the disasters that had befallen the cherished 'wooden walls of England,' that at last the closed doors of the Admiralty were opened to him. The lists he furnished were appalling indeed, and enough to impress any minds and open any doors. It was shown by Mr. Snow Harris, from carefully compiled records, based upon official documents, that in the course of forty years—from 1793 to 1832—over 250 ships had suffered from lightning. In 150 cases, the majority of which occurred between the years 1799 and 1815, about 100 main-masts of line-of-battle ships and frigates, with a still larger number of topmasts and smaller spars, together with an immense quantity of stores, were destroyed by lightning. One ship in eight was set on fire in some part of the rigging or sails, and over 200 seamen were either killed or severely disabled. But, formidable as was this account of damage done by lightning, it by no means completed the list of casualties. Mr. Snow Harris gave it as his opinion, on the authority of a great many naval officers with whom he came into contact at Plymouth, that many ships reported officially as 'missing' had been struck by lightning and gone to the bottom, with nobody left behind to tell the tale. Thus, from a reference to the log of the line-of-battle ship the 'Lacedæmonian,' under the command of Admiral Jackson, it appeared that this man-of-war sailed alongside a frigate, the 'Peacock,' on the coast of Georgia, in the summer of 1814, and that the latter suddenly disappeared in a storm of lightning, leaving no trace behind. Again, the 'Loup Cervier,' another man-of-war, was last seen off Charlestown, in America, on the evening of a severe thunderstorm, and never heard of again. A famous ship, the 'Resistance,' of forty-four guns, was struck by lightning in the Straits of Malacca, and the powder-magazine blowing up, it went to the bottom, only three of the crew reaching the shore, picked up by a passing Malay boat. But for these few survivors, Mr. Snow Harris justly remarked, nothing would have been known of the fate of the vessel, which would have been simply reported as 'missing' in the Admiralty lists. It was scarcely to be wondered at that the recital of all these tales of disasters, which might have been prevented by the most ordinary foresight in applying known means of protection against lightning, considerably excited the public mind, so that at last the Government was compelled to act in the direction into which it was impelled by the energetic Plymouth doctor. It was thus that at last, in 1839, the naval commission already referred to was appointed to give counsel as to 'applying lightning conductors to Her Majesty's ships.'

Perhaps even this step in advance might not have favoured much the cause pleaded by Mr. Snow Harris, had he not had the good fortune of finding a powerful patron in Sir George Cockburn, one of the Lords of the Admiralty. Sir George, born in London, of Scottish parents, in 1772, had all his life long taken a great interest in scientific pursuits; and the application of conductors especially had interested him much, as he had himself been a witness to frequent damage done to ships under his command by lightning. The 'Minerva,' of which he was captain at the blockade of Leghorn, in 1796, had been so struck, and likewise two ships of the flotilla, reducing the French island of Martinique, in 1809, under his direction. Having taken a prominent part in the American War of 1813–14, especially the capture of Washington, Sir George Cockburn retired from active service, and in 1818 was made one of the Lords Commissioners of the Admiralty, immediately after being returned a Member of Parliament for Portsmouth. He henceforth devoted himself more than ever to scientific studies; and, having been elected a Fellow of the Royal Society, got into acquaintance with many of its members, among them with Mr. Snow Harris, whom he came to like on account of his fervid enthusiasm in the cause he was advocating. The acquaintance proved of the highest advantage to the young Plymouth electrician. Before even the naval commission, nominated to give counsel upon the subject of lightning conductors, had given in its report, he was allowed to make trial, on board of several men-of-war, with a system designed by himself, and for which he had taken out a patent. It was not long afterwards that it was officially adopted for all the vessels of the Royal Navy, with, it is needless to say, the greatest pecuniary advantages to the designer.

The system of Mr. Snow Harris for protecting ships against lightning was similar to that suggested by Mr. Henly in 1774. Instead of hanging dangling chains from the top of the rigging into the water, he nailed on to the masts and down to the keel, slightly inlaid in the wood, a double set of copper plates, overlying each other in such a manner that the ends of one set were touched by the middle of the other. The plates were four feet in length, two to five inches wide, and one-eighth of an inch thick; they had holes drilled in them at distances of six inches apart, and were secured to the masts and further down by short copper nails. In order to prevent any break in the conductor at the junction of the successive masts, a copper plate was led over the cap, and the continuity preserved at all times by means of a copper hinge or tumbler which fell against the conductor. It was an altogether unobjectionable plan for securing protection against lightning, except that it was liable to fail under imperfect execution. Bad workmanship necessarily was fatal to it. The numerous copper plates had to be very neatly and carefully fastened together to ensure metallic continuity, in the absence of which the electric force might leave the path traced for it, diverging into neighbouring metallic masses, numerous on board ships, such as chains and anchors. It was a most costly system from beginning to end; but as it was, and, for the short time it remained in use, it accomplished all that was desired. Not one of the ships fitted with the conductors designed by Mr. Snow Harris was damaged by lightning, although many were struck, the electric spark in several cases being so powerful as to melt the too fine metal points on the top of the masts. However, the new lightning conductors had not to stand the ordeal of practice for any length of time. One by one the great wooden ships of war, once the pride and glory of England, went into peaceful retirement, to be replaced by iron machines, propelled by steam, metalled from the top of the masts to the water's edge. It had been one of the recommendations of Mr. Snow Harris to the Admiralty that his copper plates, though expensive at first, would always be worth their money as old metal; and the irony of fate would have it that the conversion of copper into silver was not to be long in waiting. Before the death of the inventor, which occurred in January 1867, his lightning conductors were fast disappearing from the ships on which they were placed. From the windows of his villa at Plymouth, Sir William Snow Harris could see a fleet of ironclads, dispensing with conductors, floating on the sea.

Notwithstanding the short use of his own special naval work which gave him fame, Sir William Snow Harris effected much in the interest of lightning protection in general. He was one of the few men in England who insisted that it was the duty of the Government, as well as of private individuals, to place lightning conductors upon all objects liable to be struck, arguing that it was little less than criminal to neglect such a simple protection against overwhelming danger. It was with some degree of vehemence, though not more perhaps than was requisite, that he stood out against those who objected to conductors because they 'attracted' lightning. Such assertion will, at the present day, be regarded as foolish by all persons possessed of the least scientific knowledge; but this was not by any means the case forty or fifty years ago, when even well-educated men denounced conductors. A civil engineer in the service of the British

Government, Mr. F. McTaggart, sent to Canada in 1826, recommended openly the pulling-down of all lightning conductors in that colony, and this too in the name of 'science,' of which he held himself to be an enlightened disciple. 'Science,' wrote Mr. McTaggart, in a book he published,[1] 'has every cause to dread the thunder-rods of Franklin; they attract destruction, and houses are safer without than with them. Were they able to carry off the fluid they have the means of attracting, then there could be no danger; but this they are by no means able to do.' Had such reasonings as these been merely the senseless talk of a few individuals, the harm done might not have been great. But it was quite otherwise. Men of power and position, if not of high education, were imbued profoundly with the same ideas as Mr. McTaggart, as evidenced in at least one striking instance, which would be scarcely credible were it not on official record. In the year 1838, the Governor-General and Council of the East India Company actually ordered that all the lightning rods should be removed from their public buildings, including the arsenals and powder magazines, throughout India. The rulers of the great country had come to their decision, as they stated, by the advice of their 'scientific officers,' who all apparently shared Mr. McTaggart's belief of the perils of 'the thunder rods of Franklin.' It was partly on the representation of the energetic vindicator of lightning conductors in Plymouth, that the order for their destruction in India was soon countermanded by the authorities in Leadenhall Street, but not before several buildings had been destroyed, among them a large magazine at Dumdum, and a corning-house at Magazine. As often before, so now, lightning itself proved the most powerful advocate of conductors, and in India they were more quickly set up than they had been thrown down.

While designing lightning conductors for the ships of the Royal Navy, Mr. William Snow Harris was called upon likewise by the Secretary of State for War to give advice as

[1] *Three Years in Canada.* 8vo. London, 1829.

to the best protection that might be given to powder magazines and other stores of war material. He did as requested, writing a very lucid paper on the subject, which met with the honour, unique in its way, of being put forward as an official document. To this day there is regularly issued with the 'Army Circulars' from the War Office a series of 'Instructions as to the Applications of Lightning Conductors for the Protection of Powder Magazines, &c.,' reproducing textually the recommendations of Mr. Snow Harris. These 'Instructions,' containing the essence of what he wrote about conductors, and, in fact, the result of all his investigations on the subject, treat the whole *ab ovo*, and as such deserve quotation. 'Thunder and lightning,' Mr. Snow Harris wrote to the War Office, 'result from the operation of a peculiar natural agency through an interval of the atmosphere contained between the surface of a certain area of clouds, and a corresponding area of the earth's surface directly opposed to the clouds. It is always to be remembered that the earth's surface and the clouds are the terminating planes of the action, and that buildings are only assailed by lightning because they are points, as it were in, or form part of, the earth's surface, in which the whole action below finally vanishes. Hence, buildings, under any circumstances, will be always open to strokes of lightning, and no human power can prevent it, whether having conductors or not, or whether having metals about them or not, as experience shows.'

Mr. Snow Harris then went on philosophising. 'Whenever,' he said, 'the peculiar agency—whatever it be—active in this operation of nature, and characterised by the general term electricity or electric fluid, is confined to substances which are found to resist its progress, such, for example, as air, glass, resinous bodies, dry wood, stones, &c., then an explosive form of action is the result, attended by such an evolution of light and heat, and by such an enormous expansive force, that the most compact and massive bodies are rent in pieces, and inflammable matter ignited. Nothing appears to stand against it: granite rocks are split open, oak and other trees of enormous size rent in shivers, and masonry of every kind frequently laid in ruins. The lower masts of ships of the line, 3 feet in diameter and 110 feet long, bound with hoops of iron half an inch thick and five inches wide, the whole weighing about 18 tons, have been in many instances torn asunder, and the hoops of iron burst open and scattered on the decks. It is, in fact, this terrible expansive power which we have to dread in cases of buildings struck by lightning, rather than the actual heat attendant on the discharge itself.'

He continued: 'When, however, the electrical agency is confined to bodies, such as the metals, and which are found to oppose but small resistance to its progress, then this violent expansive or disruptive action is either greatly reduced or avoided altogether; the explosive form of action we term lightning vanishes, and becomes, as it were, transformed into a sort of continuous current action of a comparatively quiescent kind, which, if the metallic substance it traverses be of certain *known* dimensions, will not be productive of any damage to the metal; if, however, it be of small capacity—as in the case of a small wire—it may become heated and fused; in this case the electrical agency, as before, is so resisted in its course as to admit of its taking on a greater or less degree of explosive and heating effect, as in the former case. It is to be here observed, that all kinds of matter oppose some resistance to the progress of what is termed the electrical discharge, but the resistance through capacious metallic bodies is comparatively so small as to admit of being neglected under ordinary circumstances; hence it is, that such bodies have been termed conductors of electricity, whilst bodies such as air, glass, &c., which are found to oppose very considerable resistance to electrical action, are placed at the opposite extremity of the scale, and termed non-conductors or insulators. The resistance of a metallic copper wire to an ordinary electrical discharge from a battery was found so small, that the shock traversed the wire at the rate of 576,000 miles in a second. The resistance, however, through a metallic line of conduction, small as it be, increases with the length, and diminishes with the area of the section of the conductor, or as the quantity of metal increases.'

After these theoretical explanations, Mr. Snow Harris went into the practical part of the business of protecting buildings, and, more especially, powder magazines and others containing explosive materials, against the effects of lightning. 'It follows,' he remarked, 'from these established facts, that if a building were metallic in all its parts, an iron magazine for example, then no damage could possibly arise to it from any stroke of lightning which has come within the experience of mankind. A man in armour is safe from damage by lightning. In fact, from the instant the electrical discharge, in breaking with disruptive and explosive violence through the resisting air, seizes upon the mass in any point of it, from that instant the explosive action vanishes, and the forces in operation are neutralised upon the terminating planes of action—viz., the surface of the earth and opposed clouds. All this plainly teaches us that, in order to guard a building effectually against damage by lightning, we must endeavour to bring the general structure, as nearly as may be, into that passive or non-resisting state it would assume, supposing the whole were a mass of metal. To this end, one or more conducting channels of copper, depending upon the magnitude and extent of the building, should be systematically applied to the walls. These conducting channels should consist either of double copper plates, united in series one over the other, as in the method of fixing such conductors to the masts of her Majesty's ships, the plates being not less than $3\frac{1}{2}$ inches wide, and of $\frac{1}{16}$th and $\frac{1}{8}$th of an inch in thickness; or the conductors may with advantage be constructed of stout copper pipe, not less than $\frac{1}{16}$th of an inch thick, and $1\frac{1}{2}$ to 2 inches in diameter; in either case the conductors should be securely fixed to the walls of the building, either by braces, or copper nails, or clamps. They should terminate in solid metal rods above, projecting freely into the air, at a moderate and convenient height above the point to which they are fixed, and below they should terminate in one or two branches leading outward about a foot under the surface of the earth; if possible, they should be connected with a spring of water or other moist ground. It would be proper, in certain dry situations, to lead out, in several directions under the ground, old iron or other metallic chains, so as to expose a large extent of metallic contact in the surface of the earth.'

A few pregnant sentences, which by themselves deserved the honour of permanently figuring in the 'Instructions' sent out by the War Office, completed the advice given by Mr. William Snow Harris in respect to the setting up of lightning conductors. 'A building,' he truly remarked, 'may be struck and damaged by lightning without having a particle of metal in its construction. If there be metals in it, however, and they happen to be in such situations as will enable them to facilitate the progress of the electrical discharge, so far as they go, then the discharge will fall on them in preference to bodies offering more resistance, but not otherwise. If metallic substances be not present, or, if present, they happen to occupy places in which they cannot be of any use in helping on the discharge in the course it wants to go, then the electricity seizes upon other bodies, which lie in that course, or which can help it, however small their power of doing so, and in this attempt such bodies are commonly, but not always, shattered in pieces.' He summed up as follows:—' The great law of the discharge is, progress between the terminating planes of action—viz., the clouds and earth—and in such line or lines as, upon the whole, offer the least mechanical impediment or resistance to this operation, just as water, falling over the side of a hill in a rain storm, picks out, or selects as it were by the force of gravity, all the little furrows or channels which lie convenient to its course, and avoids those which do not. If in the case of lightning you provide, through the instrumentality of efficient conductors, a free and uninterrupted course for the electrical discharge, then it will follow that course without damage to the general structure; if you do not, then this irresistible agency will find a course for itself through the edifice in some line or lines of least resistance to it, and will shake all imperfect conducting matter in pieces in doing so. Moreover, it is to be especially remarked in this case, that the damage ensues, not where the metals are, but where they ceased to be continued; the more metal in a building, therefore, the better, more especially when connected by an uninterrupted circuit with any medium of communication with the earth.'

'Such is, in fact,' he concluded, 'the great condition to be satisfied in the application of lightning conductors, which is virtually nothing more than the perfecting a line or lines of small resistance in given directions, less than the resistance in any other lines in the building, which can be assigned in any other direction, and in which, by a law of nature, the electrical agency will move in preference to any others. The popular objections to lightning conductors on the ground that they invite lightning to the building, that we do not know the quantity of electricity in the clouds, and that hence they may cause destruction, are now quite untenable, and have only arisen out of a want of knowledge of the nature of electrical action. What should we think of a person objecting to the use of gutters and rain-pipes for a house, on the ground of their attracting or inviting a flow of water upon the building; and since we do not know the amount of rain in the clouds, it is possible that the building may be thereby inundated,—yet such is virtually the argument against lightning conductors.'

Mr. Snow Harris, as already mentioned, received the honour of knighthood in 1847; and after this date lived in comparative retirement for twenty years at his residence, Windsor Villas, Plymouth. However, he was called upon, in 1855, to undertake one more important work in designing a perfect system of lightning conductors for the new

Houses of Parliament at Westminster. It was on the initiative of Sir Charles Barry, the architect, that the proposal was made by the Board of Works to Sir William Snow Harris, who accepted it with all his old eagerness for serving the cause of lightning protection. Accordingly, he drew up a plan, which he himself characterised, in a letter to the President of the Board of Works, dated February 14, 1855, as 'somewhat costly,' but which he felt sure would be absolutely certain 'for insuring the safety of the buildings against one of the most terribly destructive elements of nature.' In its essence, the plan consisted in protecting all the most elevated parts of the Houses of Parliament, including the towers, by 'a capacious metallic conductor of copper tube, two inches in diameter, and not less than one-eighth of an inch in thickness,' to be fastened together 'by solid screw plugs and coupling pieces,' and 'secured to the masonry by efficient metallic staples.' To do this, Sir William Snow Harris calculated, would involve an expenditure of somewhat over 2,000*l*., but nothing less would accomplish it. 'What I have recommended,' he wound up his letter, 'has been the result of very serious and attentive deliberation, and I conscientiously think that what I have proposed is absolutely requisite to a permanent and satisfactory security of the buildings against the destructive agency of lightning.' The Board of Works entirely adopted all the recommendations of Sir William Snow Harris, and, in accordance with them, there was included in the Civil Service Estimates laid before the House of Commons in the session of 1855 a vote of 2,314*l*., on account of 'works necessary for securing the new Houses of Parliament against danger from lightning.'

The vote passed without demur. It was in the height of the Crimean War fever, political questions absorbing all others. Perhaps in a time of less excitement some voice might have been raised in the House of Commons asking whether it was wise to spend over 2,000*l*. in putting up lightning conductors, without previously ascertaining, from the best scientific authorities, that the system adopted was the best, and absolutely efficacious. The strongly recommended 'copper tubes,' with their 'screw plugs and coupling pieces,' were at least a novelty, not having stood the test of experience, and there were practical men who shook their heads when they heard of them. However, with war discussions raging fiercely, and reports of battles and sieges absorbing all attention, the House of Commons had no time to bestow upon such trifling matters as that involved in the plans of Sir William Snow Harris; and thus the vote passed unchallenged. Perhaps silent repentance came afterwards to the official mind. At any rate, as it was the first, so it was the last time of Parliament granting money for lightning conductors.

CHAPTER IX.

THE BEST MATERIAL FOR CONDUCTORS.

'THE art of protection against lightning,' says a recent German writer, in a book on conductors, 'is precisely the same now as it was a hundred years ago: still, it has made immense progress since that time.' Though apparently involving a paradox, the words nevertheless are literally true. The art, or rather science, of guarding objects against the destructive effects of lightning is theoretically the same as it was in the days of Benjamin Franklin; nevertheless, the practical execution of the appliances necessary to attain this aim has undergone extraordinary improvements since that time. This has been due simply to the astounding progress of the metallurgical arts for the last forty or fifty years. With the help of machinery on a colossal scale, such as was never dreamt of before, our factories have come to produce metallic masses of dimensions and shapes such as make all former achievements of the kind appear utterly insignificant. We build huge iron ships, armed with cannon of ponderous weight; we throw iron bridges across rivers and arms of the sea; we lay metallic cables through the ocean and over the earth, encircling the globe. All these wonderful achievements, in which the development of engineering science went hand in hand with that of tool-making and the ever-growing employment of the power of steam, have gone to the constant improvement of lightning conductors. They have benefited, indirectly, in the result of great inventions, and of immense toil and labour, originally directed to other ends.

THE BEST MATERIAL FOR CONDUCTORS.

There is something half touching, half comical, in reading of the troubles which Benjamin Franklin had to undergo before he was able to set up his first lightning conductor. He could meet with no assistance but that of the blacksmith of little Philadelphia; and the ability of the latter in the art of forging iron rods more than a few feet in length was of the most limited kind. The ingenuity of Franklin overcame this difficulty by a variety of clever contrivances, such as connecting a number of small rods by caps and joints, fitting closely; but others were not so successful as he in the matter. Even in Paris there were no artisans to be found, for many years after lightning conductors were first recommended, able to make them, and foreigners, chiefly English, had to be brought there for the purpose. The difficulties arising from this backward state of the industrial arts were greatly increased by the belief, prevalent for a long time, that lightning conductors, to be efficient, ought to be of very great height, their so-called 'area of protection' being in proportion to their height. The supposition, originating in France, was carried to extremes in that country, chiefly through the teachings of M. J. B. Le Roy, a very able but eccentric man. Guided by vague analogies in electrical phenomena, M. Le Roy, who enjoyed in his time—the latter part of the eighteenth century—the reputation of being an authority on the subject of lightning conductors, laid it down as an indisputable fact that the 'Franklin rods' only protected buildings if rising high above them. He recommended the length of the rods above the chimney, or summit of any edifice, to be not less than fifteen feet, guaranteeing that, if of this height, they would offer absolute protection against lightning over an area of four times the same diameter—that is, sixty feet. Modern experience has proved this to be an absurdity; still, in the infancy of all knowledge about lightning conductors it was, perhaps, not unnatural that even learned men should believe in such fancies. Lightning was looked upon, not only in name but in reality, as an electric 'fluid,' and the conductor was supposed to draw this 'fluid' from the clouds. Therefore it was but cogent reasoning to raise conductors as high above the roofs, and as near to the storm-clouds, as could possibly be done. If possessed of modern means for manufacturing pieces of metal of almost any length, M. Le Roy would not improbably have recommended to elevate lightning conductors a couple of hundred feet, instead of only fifteen, above the summit of buildings.

It was owing chiefly to the difficulty of forging long iron pieces, and of welding them together in a satisfactory manner, that, for many years after lightning conductors had been introduced into Europe, there were constant attempts made to find substitutes for the rods devised by Franklin. Chains were largely used towards the end of the last and the beginning of the present century, both in France and Germany, their employment having been suggested by the example of the English navy, where they were introduced, as already mentioned, upon the recommendation of Dr. Watson. The Continental mode of using iron chains for the protection of buildings against lightning was to hang them between the upper part of the conductor, surmounting the roof, which continued to be a straight piece of metal or rod, and the lower portion buried in the ground, sometimes, but not always, likewise a chain, but thicker than the rest. The characteristic of this method, and showing its long existence, is that it gave rise to a nomenclature existing to this day in France and Germany, where in all books on lightning conductors they are described as consisting of three distinct parts. The French call the upper part of the rod, over the roof, '*la tige*,' the stem or stalk; and the Germans, '*die Auffangstange*,' literally the reception-rod. In both languages the middle part, from the roof downwards to the earth's surface, is described as the conductor proper, '*le conducteur*' and '*der Leiter*.' Again, the lowest underground part of the conductor is designated, by the French, '*la racine*,' the root, and by the Germans as '*der Bodenleiter*,' or the ground-conductor. It has often been said that, as language springs from ideas, so it reacts upon them, and if the proposition be true, as most will admit, the French and German designations of the parts of lightning conductors—also to be found in Italian, and adopted in a few of the older English treatises on the subject, mostly translations—have a strongly misleading tendency. Nothing could be further from the truth than the assertion that a conductor ought to consist of three distinct parts. On the contrary, the more it is 'one and undivided,' the better it will be as a lightning protector.

The use of iron chains as conductors gave rise to very many fatal accidents, and for a time resulted in an outcry that the system itself could not be depended upon, as it was known to be not always efficacious. Lists were published of numerous instances in which buildings with what were supposed to be the best conductors were struck by lightning, from which it was argued that Franklin's great discovery of the electric force always seeking a metallic path to the earth was a myth. It was not till some painstaking scientific men, deeply interested in the subject, had set to work to discover the causes of the failure, that the whole became plain enough. The chains, in some of the instances in which they had proved inefficient lightning conductors, were found to be corroded to such an extent as barely to hang together. Of course this corrosion would impair the efficiency of the conductor by reducing the quantity of metal; but the chief objection to the use of chains lies in the fact long ago pointed out by Mr. Newall, that even supposing a chain were formed of links of half-inch copper rods, and were perfectly bright and clean, the area of the conductor is reduced to a mere point where the links touch each other, and the resistance becomes so great in such a small conductor that instances have been recorded of the fusion of the links. In other cases, as in that of H.M.S. 'Ætna' in 1830, the chain was boomed out, and did not touch the water!

Simultaneously with the chains, there was trial made, in

several Continental states, and also in England, of several other metallic conductors besides iron. Tin and lead had both their advocates, but the latter more than the former, on account of its far lower price. As regards tin, it had really no advantages whatever over iron, except pliability and non-oxidation. Against this was to be set that it was much more expensive than iron, with only about the same conducting power, according to Becquerel, Ohm, and other investigators. Professor Lenz, it is true, ranked tin very much higher, asserting, from experiments of his own, that its power of conductivity was nearly twice that of iron; and it was partly owing to his great influence that the metal obtained a trial in several countries, more particularly in Russia and in the United States of America. Still, the result was not satisfactory on many accounts, and its price alone brought tin to be soon abandoned as a conductor. Lead had a far longer trial. Its cheapness recommended it strongly, and equally so its extreme pliability. One of the greatest difficulties of the constructors of 'Franklin rods,' when first they came into demand, was to make the iron pieces fit properly around sharp corners of buildings, either by bending them in fire, or, as was more commonly done, soldering them together, or employing screws and other joints. But it was early discovered that these junctions, when occurring at acute angles, were bad conductors, occasioning sometimes the electric force to leave its traced course, and fly off in some other direction. It is probable that, in several well-authenticated instances in which this really did happen, the joints were eaten away by oxidation, as in the case of the chains; still, the effect of such occurrences was all the same. The joining of strips of lead together was a far easier task than that of handling iron in the same way, particularly for inexperienced workmen, and thus the employment of the metal continued for some time. However, it had to be abandoned gradually, on account of its manifest disadvantages. Its extreme softness, which made it liable to be broken by any accident, was one of them, and, still more so, its want of conducting power—only about one-half that of iron. Thus leaden conductors slowly went out of use, except in the form in which they still act often to great advantage, that of water-pipes.

Among all the experiments made for producing the most perfect lightning conductors, the one which created the greatest attention, some fifty years ago, both on the Continent and in England, was the employment of ropes made of brass wire. They were first recommended about the year 1815 by a professor at the University of Munich, J. C. von Yelin, distinguished for his researches into the nature of thunderstorms. Through his influence most of the public edifices of Bavaria, more particularly the churches, were provided with conductors of brass ropes; and within a few years their employment became so popular, owing to the ease with which they could be attached to all buildings, that even the Roman Catholic clergy changed their attitude, and, from being opposed to 'heretical rods,' advocated their extension in every direction. But it was not long before the trust in brass ropes as protectors against lightning was rudely shaken. Several instances occurred in which buildings so protected were struck and damaged by lightning, and at last there came a case which attracted the widest attention, leading, on account of its supposed importance, to the institution of a Royal Commission to report thereon. The little town of Rosstall, in Franconia, Bavaria, had a church the steeple of which was 156 feet high; and, standing on the brow of a hill, it overlooked the country far and wide, visible for many miles. Necessarily much exposed to the influence of lightning clouds, it had been provided with one of the best brass-wire conductors, designed by Professor von Yelin himself, and made of unusual thickness, being over an inch in diameter. Nevertheless, on the evening of April 30, 1822, while a dark storm-cloud, of extraordinary thickness, was passing over Rosstall, a heavy flash of lightning was seen to fall vertically upon the church steeple, followed by a terrible peal of thunder, which seemed to shake the earth. When people looked up they beheld the church clock thrown from its place, and part of a lower wall of the edifice thrown to the ground. It was clear that the electric discharge from the atmosphere had been one of unusual energy, but equally clear that the trusted conductor had not done its work.

It was partly through scientific controversies about the relative conducting value of metals, and partly through the action then taken by several German Governments of providing all buildings with lightning conductors, that the Rosstall case excited an extraordinary interest at the time. The Royal Commission appointed by the King of Bavaria, presided over by an eminent savant, Professor Kastner, went to Rosstall to inspect the effects of the lightning discharge, and Professor von Yelin did the same, as an independent, though not disinterested witness. Their reports as to actual facts were the same. The lightning, after striking the steeple of the church, had melted the top of the '*Auffangstange*,' or highest part of the conductor, and further down had passed along the brass rope till coming to the clock, only a few inches distance from it. Here the electric force had evidently divided itself into several streams—the one exerting its disastrous effects upon the clock and brickwork, and several metallic objects underneath, and the other passing down the rope conductor, but not without bending it, and, in one or two places, tearing it to pieces. Such were the facts, visible to all eyes. But the conclusion drawn therefrom differed widely. The members of the Royal Commission made it public that the reason of the Rosstall lightning conductor not having been efficient had simply arisen from its nature. Brass-wire ropes, they declared, though perhaps useful against small discharges of electricity, formed no reliable safeguards against powerful ones; and they therefore strongly advised a return to the old-fashioned iron rods. The conclusion was vehemently disputed by Professor von Yelin. He admitted that it might be better, to provide for the proper discharge of extraordinary masses

THE BEST MATERIAL FOR CONDUCTORS.

of the electric force, to make his brass ropes, when applied to high churches and other large edifices, even thicker than they had been at Rosstall; but at the same time he utterly denied that, even in this case, they had been the origin of the disaster. He showed that the real cause of it was that the conductor had not been laid deep enough into the ground, so as to touch moist earth. The church stood upon sandy soil, on an eminence, and to touch 'good earth' the brass rope ought to have been sunk down to a depth of at least fifty feet, whereas it did not reach one-third of that depth. The professor was undoubtedly right, but his antagonists nevertheless prevailed. A public prejudice, which no argument could overcome, set in against brass-wire conductors, and they were pulled down from nearly all buildings on which they had been laid, to be replaced by iron rods. Some time had to elapse before real justice was done to metallic ropes as lightning conductors.

With our present knowledge of electrical phenomena, and the practical art of making conductors, it may safely be affirmed that the Munich professor was right in recommending ropes, though not in approving of brass as the best metal. In its very nature, brass, a compound, can never be thoroughly reliable, because its conducting power varies according to its composition. The facility with which it allows the electric force to pass through it depends, in fact, entirely on the amount of copper which brass contains, and is greater or less accordingly, since the other metal entering into its composition, zinc, has less than one-third the same conductivity. Now brass is, for various purposes, made sometimes of 70 parts of copper and 30 parts of zinc, and again, only equal amounts of both metals, setting calculation as to its conducting power entirely at nought. But besides this, brass has the great fault of being excessively liable to destruction by atmospheric influences, and it was found, among others, in Germany, that while brass ropes were used as lightning conductors, they were frequently destroyed, in a comparatively short space of time, by the action of smoke alone. It is true, the Continental mode, existing both in France and Germany, of spanning conductors over the tops of chimneys—illustrated in the engraving

here as a warning 'how *not* to do it'—had much to answer for this atmospheric deterioration, since even tougher metals than brass could not be expected to stand the constant action of smoke, often containing sulphurous fumes. But even without such an evidently absurd arrangement as that of running any conductors, whether in the form of ropes or cords, across the orifices of chimneys, brass could never have answered all the requirements of a lightning conductor. It was with justice that brass-wire ropes were nearly altogether discarded some thirty or forty years ago, after having had a short-lived reputation.

That copper should not have been employed, long before brass and other metals, in serving mainly for lightning conductors, its pre-eminence for this purpose being undisputed, would seem a strange fact, were it not explicable on several grounds. The first was the cost of the metal, which, though varying in price, is seldom less than six or seven times that of iron. It was needless for the advocates of copper as conductors—and there were not a few from the time its high conductive power had been demonstrated—to say that if copper was six times as dear as iron, it was likewise six times better as a carrier of the electric force, and that consequently the price, in respect of applicability for lightning protection, was in reality the same. But the reply to this was that copper, being one of the most expensive metals, except the so-called 'precious' ones, was exposed to the temptation of theft, and ought therefore not to be employed, since it was possible that vagrants, or other people, might tear off at any time the, in more than one sense, valuable pieces of metal protecting buildings against destruction from lightning. The argument, perhaps, was not worth much, but a better one not mentioned was in the background. It was, till quite recent times, an achievement of the greatest difficulty to manufacture long rods or bands of sufficiently pure copper to serve as lightning conductors. Sir William Snow Harris attempted, as already related, to get over this impediment by taking short plates, and fastening them together, and over each other, by copper nails. But this process, besides being enormously expensive, was in many other respects unsatisfactory, notably in that it made a shifting of the plates possible, and by the destruction of a few of them ruined the whole system. The pith and substance of the whole was the technical difficulty of hammering or drawing pure copper out into great lengths. That it must be pure was essential, the fact being thoroughly established that the electric conductivity of copper, mixed with impurities such as arsenic, is often not two-thirds, and sometimes not as much as one-half, that of the pure metal. This was conclusively shown by Sir William Thomson in a series of researches, and likewise by that distinguished investigator in the conductivity of metals, Professor Matthiessen. The latter, while placing copper on the same rank with silver, and far above gold—100 to 78—furnished the following instructive list as to the relative value of different kinds of copper :—

Pure copper	100·00
Best American copper	92·57
Australian copper	88·86
Russian copper	59·34
Spanish 'Rio Tinto' copper	14·24

It will be seen that, according to the investigations of Professor Matthiessen, admitted on all hands to be correct, the copper lowest in the list, the 'Rio Tinto,' is barely equal to iron in electrical conductivity, and, not having the hardness of the latter metal, would be in every way inferior to it as a lightning protector. The employment of the purest copper therefore became an essential point in the manufacture of lightning conductors.

Fortunately, the difficulty was solved, at an earlier period than might have been expected, by the demand for submarine cables. These had to be made of wires of the highest possible electrical conductivity, and the matter being one of high financial and commercial importance, manufacturers soon began to use the utmost care in selecting ores containing the smallest amount of metallic impurities. We believe the lightning conductors now manufactured at the extensive works of Mr. R. S. Newall, F.R.S., established at Gateshead on Tyne about forty years ago, have generally a conductivity of 93 per cent. of pure copper. It was laid down by one of the most eminent scientific men of the day, not long ago, that the three principal qualities of a good lightning conductor ought to be a maximum of conductivity, of durability, and of flexibility that could be obtained, and there is nothing coming up to this standard so well as ropes of pure copper.

CHAPTER X.

HOTEL DE VILLE, BRUSSELS, AND WESTMINSTER PALACE.

THE systems of lightning-conductors used for the protection of the Hôtel de Ville and Westminster Palace seem worthy of separate description, as showing the methods employed by Professor Melsens and the late Sir William Snow Harris, both eminent authorities in their respective countries. The two buildings are so entirely distinct in their character, that it will be seen at once that very different methods had to be employed in rendering them safe from the effects of thunderstorms.

The Hôtel de Ville, Brussels, one of the finest Gothic structures in the Netherlands, is fitted with an elaborate system of lightning-conductors, erected under the superintendence of Professor Melsens, a distinguished electrician and scientist. He has for many years advocated the method of employing a great number of small lightning-rods, in preference to one rod of large size, for the protection of buildings from the effects of lightning; the main characteristic of his system being that of covering the building with a network of metal furnished with very many points, combined with numerous and ample earth-contacts. This idea has been thoroughly worked out at the Hotel de Ville, Brussels; and probably no other building is so completely guarded from the dangers of thunderstorms. The principal feature of the Hôtel is a large central building, with a pinnacled turret, from which rises a lofty spire, nearly three hundred feet high, and adorned with four galleries, each with corner pinnacles. Upon the top of this spire is a gilded colossal figure, seventeen feet high, of St. Michael, holding a naked sword and standing upon a dragon. This acts as a vane, and the point of the sword forms the highest terminal conductor of the system. The main block of the Hôtel is ornamented with six turrets, from each of which springs a small spire. In the rear is a courtyard formed by buildings annexed to the front main block, and composing the remaining three sides of this inner quadrangle.

The figure of St. Michael, all the parts of which are rivetted and soldered together, rests on a pivot of iron, three and a half inches in diameter, which is deeply embedded in the stonework of the spire. The weight of the vane produces a metallic connection with the pivot, and the top of the platform in which the pivot is fixed is covered with sheet-copper. Around this and in connection with the pivot are fixed eight perpendicular galvanised iron conductors, two-fifths of an inch in diameter, and provided with five points each. A flash of lightning striking the statue would thus reach the pivot and then be divided between the eight conductors. Just below the platform are placed, at an angle of 45 degrees, eight large points six and a half feet long. These are fastened to an iron band which encircles the spire, and are connected with the eight conductors by means of a mass of zinc. Thus the pivot of the statue, and consequently the statue itself, the eight conductors, the eight large points, and the forty small points on the conductors, constitute a protection which dominates the edifice, and represents a circular space of about five and a half yards in diameter; that is, between the extremities of the large points which project from under the platform. In this manner a flash of lightning is instantly distributed and conveyed by the conductors to the ground. It may be mentioned here that a thin copper wire, insulated by three coatings, is fixed on the north side of the iron band in which the large points are fastened;

the other end of this wire is left free, and can be utilised as a conductor for a rheometer or any other electric machine which it might be thought proper to use permanently for the registration of lightning striking the conductors.

The eight conductors have each an unbroken continuity of about 310 feet; and they collectively show a continuous section of nearly one inch—almost half as much again as the limit of safety given in the 'Instruction' of the Paris Academy. Although, in Professor Melsen's opinion, rods of somewhat less diameter would have been amply sufficient for security, he chose the largest size which could be easily bent to the varying contours of the building, and also as allowing for the expansion and contraction caused by changes of temperature. If conductors of only one quarter of an inch diameter had been used they would, it is true, have shown a total section just above the limit of the 'Instruction;' but, since Coulomb has demonstrated that tensional electricity is more particularly carried on the surface of bodies, M. Melsens thinks it is necessary to consider the action that this surface might exercise in the easy transmission of electricity. Some old German writers on this subject went so far as to assert that the conductivity was proportioned to this surface. They therefore recommended flat bands or hollow tubes in place of rods. Although exact figures cannot be given as to the effect due to the area of the surface, M. Melsens considers that it is unquestionable that the relation of the section to the surface has a marked and definite, although at present unknown, result. In the case of the Hôtel de Ville, Brussels, he thinks the eight conductors possess a signal advantage over one conductor, even though it had a larger section —say one inch. Experience will doubtless teach how to determine more precisely the extent of this surface-action.

The eight conductors descend the length of the octagon of the spire until they reach the first gallery; going round this they pass over the balustrade, and then converge towards each other; are carried over a prominence in the roof; and as they pass along gather up other conductors of similar size from the ridges and parapets of the buildings which form the quadrangle. Projecting vertically from these horizontal lengths of the conductors are a large number of points and aigrettes. The summits of the lower tower are also furnished with a great many points. These eight main conductors are then taken down the wall of the building into the courtyard, and at about three feet from the ground are carried into a box constructed of galvanised iron, and in it are connected into one solid mass by zinc, which has been poured molten into the box. Almost throughout their length the conductors are left loose, so as to remove all complication arising from dilatation; the play of this dilatation being rendered easy on account of the small section of the conductors, which bend readily.

In accounts of lightning striking buildings which have been provided with lightning-conductors, it is almost invariably found that these conductors are incomplete, and have generally been fixed by persons ignorant of the scientific questions involved. When the facts in such cases are carefully examined it is found, as a rule, that the defect is in the connection with the water underground, or in the bad conductivity of the earth in which the conductors terminate. In establishing a perfect communication with the earth, M. Melsens considers it is necessary, not only to place the conductors in contact with water, but also to see that the contact extends over a large surface. The Paris Academy 'Instruction' recommends this precaution, but in a very vague and too succinct a manner. To the above rule may be added another condition, namely, that the earth-connection should be large in proportion as the site of the building is redundant in metal products in direct or indirect contact with the ground, the subsoil, or the damp earth of the foundations, and sometimes even with water itself. With regard to the metal contained in the materials of buildings, it is not sufficient to establish a connection at one point only, as is generally supposed. On the contrary, it is important that all the metal-work should be connected with the conductor at least at two points, in order to realise closed metallic circuits, and thus offer an entry and exit, or a free metallic course, for the current of electricity. The foregoing statements have been placed here chiefly because the principles they convey have been so rigidly, and at the same time successfully, carried out by Professor Melsens at the Hôtel de Ville, Brussels.

To return to the eight conductors and the earth-connections provided for them. It has been shown that these con

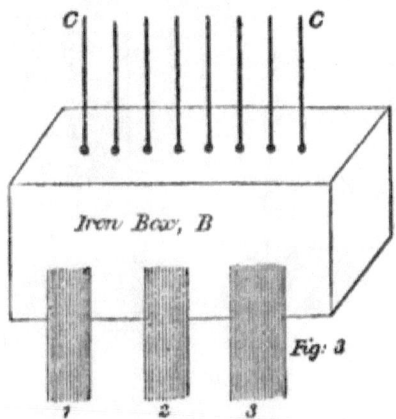

ductors, after descending the wall of the building, reach a point about three feet from the ground, where they are embedded in a rectangular box of galvanised iron, which is eight inches long, three inches broad, and three and a half inches high. In the bottom of the box are three holes, through which pass three series of eight conductors, each series being of the same diameter as those which descend from above; the conductivity being thus increased threefold. All of these are formed into one mass by the zinc, which has been poured into the box in a molten state, so that they constitute with the eight rods from above, one integral conducting system. In the illustration which is here given the box is represented by B, and the eight main conductors coming down from the building by c c. The three series of rods numbered 1, 2, 3 show the triplicated conductors issuing from the box. The first series is placed in communication with the water by means of an iron pipe, which carries it underground to a well. Here the rods are inserted in a large tube six and a half feet long and nearly two feet in diameter (see engraving). This tube is let down

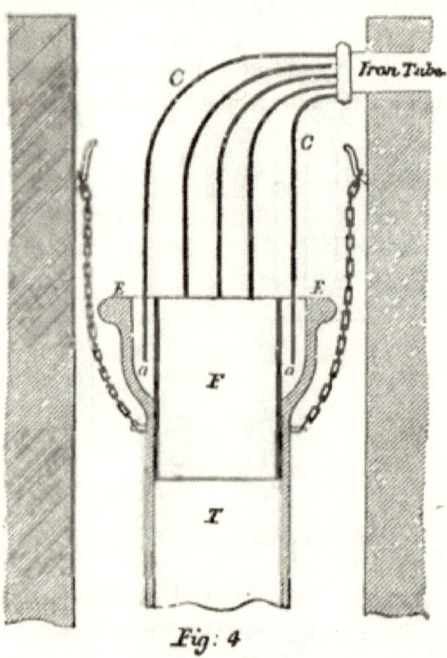

Fig. 4

almost four feet below the level of the earth, and sustained by two chains hung on two iron holdfasts fixed in the side. The conductors c c are fastened to this tube in the following manner:—A small length of straight iron cylinder is placed outside the flange of the tube; and the ends of the conductors being arranged between the cylinder and the flange, the space $a\ a$ is filled with molten zinc; thus rendering the substance of the iron tube and that of the conductors metallically continuous. The well into which the tube is sunk furnishes perpetually a contact of eleven square yards between the water and the iron of the tube. Into the space $a\ a$ is also introduced a large number of small galvanised iron wires to act as auxiliary conductors; these are terminated by being brought to a point and soldered to the mass of zinc. In order to prevent as far as possible the formation of rust, a large quantity of lime is thrown into the well, in order to make the water alkaline. The second series of conductors, painted with coal-tar, is placed in a covered metal gutter and carried some distance to a gas-main in a spot where the earth is moist. The conductors are fixed by means of a large copper plate, which is soldered to the gas-pipe or main. On the copper plate are fastened sixteen large-headed brass screws, to which the conductors are secured. This arrangement is enclosed in brickwork, the wires being painted with coal-tar; and a quantity of boiling tar is poured on the copper plate, over which is laid a cloth, thus preserving the whole from oxidation. The third series of conductors is carried in a gutter, similar to that which contains the second series, to a water-pipe in the Place de l'Hôtel de Ville, and the wires are fixed to it in the same way.

It may be added that the whole of the conductors above-ground—with the exception of the points—are painted with oil.

Although it is correct that the coke generally placed around the earth-connection of conductors aids by its good conductivity to bring them in contact with a large surface of earth, Professor Melsens has preferred to employ tar, which, it is true, is insulating, but helps materially to preserve the conductors. It is estimated that the entire contact between the earth and the underground surface of iron is about 300,000 square yards.

Professor Melsens thinks it is worthy of note that, although copper is a better conductor of electricity than iron, it has less molecular strength. Where thin iron wire would simply be beaded—without losing its conductivity—

by an exceptionally strong charge of electricity, copper wire of the same thickness would by a similar charge be dissipated to a black powder. Professor Melsens has verified this in some very interesting experiments. The large use of iron in his system of conductors on the Hôtel de Ville, Brussels, was rendered imperative by reason of the enormous cost of sufficient copper for such an extensive system. But Professor Melsen's experiments, nevertheless, give some support to the selection of iron for large and complete works of this kind.

Sir William Snow Harris, in his arrangements for the protection of the Palace of Westminister from lightning, has endeavoured to perfect the general conductivity of the whole mass of the building, and so make it assume the same relation to the electric discharge as if it were a complete mass of metal. Westminster Palace differs in one important respect from the Brussels Hôtel de Ville—the general level of the roofs is covered with iron coated with zinc, and in many places directly connected with the earth by cast-iron water-pipes. The roofing thus constitutes, although imperfectly, and only to a limited extent, a protection of itself. Sir William Snow Harris had, therefore, chiefly to provide for those portions of the building which are above the general level of the roofs, and, by the use of ample conductors of copper, to make up for the comparatively low conductivity of the roofing and the iron pipes which connect it with the earth.

From the terminal which forms the highest point of the large central tower is brought a copper tube of two inches diameter and one-eighth of an inch in thickness, the joints of which are secured by solid screw-plugs and coupling-pieces. This tube is carried down in the south-west angle of the tower and fastened to the masonry by metallic staples. At the junction of the tower with the roofs the tubing is, or at any rate was, thoroughly connected with the metal of the roof, and then continued to the earth in as straight a course as practicable, and there terminates in two projecting branches made of solid copper rod. By carrying this copper tubing direct to the earth, instead of terminating it in the metal-work of the roof, the electrical discharge obtains a conducting medium of the same power throughout, in place of having to leave a high power for one of lower conductivity.

The Victoria and Clock Towers, which are each 300 feet high, are both fitted with a copper band, five inches wide and a quarter of an inch thick. These run down the walls, and are connected with the metal of the roof, and also with the metallic rail of the staircase within each tower. The ornamental turrets and pinnacles of St. Stephen's Porch are protected by small bands of sheet-copper, two inches wide and one-eighth of an inch thick; these are also placed in connection with the metal of the roof.

The north and south towers of the central block, and the north and south wing towers of the front facing the Thames, have attached to them bands of sheet-copper running from their respective vanes to the roofing below. The bands are connected with the metal of the roof, and are then carried down independently to the earth, in a similar manner to that adopted on the large central tower.

The only other prominent portion of the edifice is the ventilating shaft of the House of Commons, where, during the sitting of Parliament, a coke-fire is generally burning, and from which, therefore, a stream of warm and rarefied air is constantly being emitted. The conductivity of an ascending column of warm vapour is known to be great, and accidents from this cause are of frequent occurrence, although very often they are not ascribed to their true source. To obviate this danger, the ventilating shaft is provided with a copper tube conductor, fixed on its eastern side, and connected with the metal of the roof.

This short description of the measures adopted by Sir William Snow Harris for the protection of Westminster Palace contains all the salient points of the system which at that time, some twenty years ago, was doubtless the best that could be devised. But, although nearly 4,000*l.* was spent upon this work, from that time to this, as far as can be ascertained, these lightning-conductors have never been tested! It is therefore very possible, and indeed probable, that on the occurrence of any very heavy thunderstorm they would be found wanting, and considerable damage would ensue, the extent of which no one can estimate.

CHAPTER XI.

WEATHERCOCKS.

ALTHOUGH such an opinion seems scarcely orthodox, it may, and not unreasonably, be doubted whether weathercocks are of any great use in demonstrating the direction of the wind. The under-currents of air are so numerous and so conflicting—especially in towns where the houses are lofty—that it is quite possible for two weathercocks at different ends of the same street to show at the same moment the wind blowing from opposite directions. However, the prevailing custom of placing these ornaments, in connection with lightning conductors, on the highest points of large buildings renders necessary some explanation of the manner in which they should be fixed; for if they are improperly or negligently attached to the lightning conductor, the continuity of the latter may be rendered defective, or at least seriously impaired.

The two main points to be kept in view are, that the weathercock should move freely with the wind, and that the continuity of the lightning conductor should be preserved. One method of obtaining this result is to put the weathercock into a circle, with the terminal rod of the lightning conductor on the top. This is called the 'nimbus cock,' and is in somewhat doubtful taste. The continuity, however, is perfect, and the cock, which is simply placed on to a point, moves easily with the wind. An example of this cock may be seen erected on the central finial of the Cathedral at Amiens.

A different way is to make the terminal rod of the conductor serve as a pivot for the cock, as shown in fig. 1. This is the usual kind of weathercock used in England, and is considered by many to be one of the best forms. It is arranged in this manner: the actual terminal of the conductor ends with a rounded or sharp point of steel, and acts as a spindle, on which the weathercock revolves. It varies in diameter from five-eighths to three-quarters of an inch. A tube, from seven-eighths of an inch to one inch in diameter, and on which is fixed the cock, is made to fit on to this terminal or spindle. This tube contains at the extremity

Fig. 1.

of its interior a piece of steel or glass, or sometimes a glass ball, and being lengthened to a point, with a platinum or copper tip, serves as the point of the lightning conductor. This weathercock is generally called a 'formed cock;' it measures at its extreme length about twenty-one inches, and weighs about twelve pounds. It will be seen that in this method there is nowhere absolute contact between the point and the pivot; consequently electric sparks must be caused by the current of electricity. Besides this defect, if the metal becomes oxidised between the surfaces, insulation will be the result. This plan, though often adopted, sacrifices the principal for the sake of the accessory.

Fig. 2 shows another method of fixing the weathercock on to the conductor. It is called a 'solid cock,' and is cut out of sheet-copper one-sixteenth of an inch thick; it revolves on a spindle in the manner shown in the engraving. This spindle, on which the cock or 'blade' works, differs in diameter according to the weight of the bird, the height and style of the building, &c., but as a rule it is from five-eighths to three-quarters of an inch in diameter.

It is usually used for Gothic buildings and private mansions, but should not be adopted, as it is apt to be lifted off its spindle by the wind. When a down-current of wind

Fig. 2.

takes place there is generally an up-current at the same time, and there is a possibility of the cock being blown off during a gale. If it is used it should have a long point fixed above it.

A better arrangement than either of the two preceding ones is to let the point go quite through the weathercock in

Fig. 3.

an encasement. The cock is then supported on a small round embasement, upon which are placed three small rollers (see fig. 3); on these the vane revolves easily, the continuity of the lightning conductor is perfect, and the weathercock freely turns round on the point so long as the small rollers are in order. In this arrangement, as in most others, the best material for the point is copper. Steel has occasionally been used, but it was found that in a very short time the rust had so eaten into the joints that the cock would not turn with the wind.

The most complete and enduring method is that indicated in figs. 4 and 5; this is the plan adopted in England for all the best work. By this arrangement, the friction being very much diminished, the weathercock revolves with great ease and freedom; the possibility of its getting out of order is

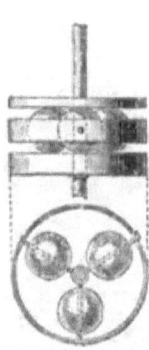

FIG. 4.   FIG. 5.

reduced to a minimum; and the continuity of the lightning conductor remains unimpaired.

It is accomplished in this way :—A circular plate, through the centre of which the point passes, is permanently fixed some distance down the point. On this circular basement rest three glass balls, rolling on three axes radiating from the centre, i.e. the point, and fixed in their outward extremities to a ring which surrounds the balls (see fig. 5). On these balls is placed another circular plate, on which is fixed the weathercock. The weathercock and circular plate, with a hole through the centre, is simply put on to the point of the conductor, and allowed to rest loosely of its own weight on the balls of glass.

CHAPTER XII.

LIGHTNING PROTECTION IN FRANCE AND AMERICA.

In this chapter it is proposed to give a brief *résumé* of the different systems of constructing, erecting, and repairing lightning conductors in France and America. The laws of electricity being the same all the world over, the methods employed in these countries are necessarily similar in their essential principles; nevertheless they vary somewhat in detail, both from each other and from the work of the best firms in England.

Until a very few years ago the lightning conductors throughout France, although many in number, were in a very neglected state. Badly constructed in many cases, their original faults had grown worse from want of attention. The connection of the terminal rod with the conductor was generally made by means of a strap or iron collar, which, after a short time, rusted to such an extent that the continuity was practically reduced to nothing, and the conductor, so far from being a protection to the building, was a positive danger to it.

Latterly, however, a reaction has taken place, and a more careful method of connecting the various joints of the conductor has been contrived, or rather, revived, and a better system of periodical inspection and testing is carried out.

Under the French system, what is called in England a lightning conductor, and to which the French give the name *Paratonnerre*, is nominally divided into three parts: the terminal rod, the conductor, and the *racine*, or root, i.e. the

earth connection. With regard to the terminal rod, the
'area of protection' theory is, in France at any rate, still
believed in by a great many people. In that country, as a
rule, it is made of wrought iron in a single length, and poly-
gonal or slightly conical; its height depends upon the size
and area of the building it protects, the general presumption
being that, under ordinary circumstances, a terminal rod
will protect effectually a cone of revolution, of which the
apex is the point of the rod, and the radius of the base a
distance equal to the height of the said rod above the ridge,
multiplied by 1·75. Thus a rod rising eight yards above
the ridge of a building would effectually protect a cone-

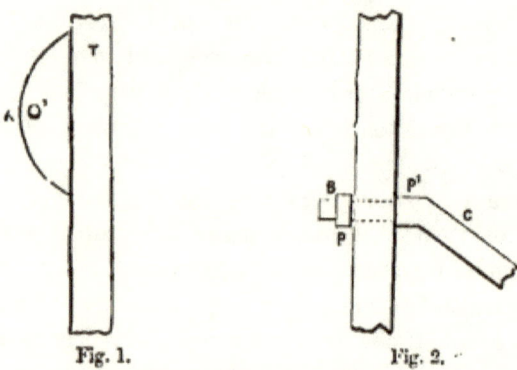

Fig. 1.   Fig. 2.

shaped space, the base of which, at the level of the ridge,
has a radius of $8 \times 1·75 = 14$ yards. In actual practice
somewhat wider limits are allowed. The height of the terminal rod having been determined according to the circumstances under which it is erected, it is then galvanised with
zinc in order to prevent oxidation, and the connection between the terminal rod and the conductor is formed by
means of the following arrangement. A little above the
base of the terminal rod, say about eight inches from the
roof of the building, a flange A (see fig. 1) is welded with a
hole pierced through it. Through this hole the conductor,
previously filed down to the proper dimensions, must be
tightly passed. After scraping the iron around the hole, a

washer of lead is placed at P and P' (see fig. 2), and the button B, by means of a strong layer of solder, thoroughly binds everything together. In this way an excellent joint is obtained; the contact surface is considerable, and, if the work is carefully done, the joint is completely preserved from rust.

The point of the terminal rod, although sometimes made of platinum, generally consists of either pure red copper, or, what is considered still better, an alloy of 835 parts silver and 165 parts copper. It is fastened to the terminal rod in the manner shown in fig. 3.

C is the trunk of a red copper cone, upon the top of which a point, P, made either of platinum or of an alloy of silver and copper, as before mentioned, is screwed, pinned, and strongly soldered with pewter solder at a, the whole being screwed on to T at b. To ensure complete contact and continuity, a washer of freshly-scraped lead is inserted between C and T, and the whole of the joint thickly covered with a layer of pewter solder. It may be added that the point forms an angle of fifteen degrees with the vertical, consequently the point terminates in an angle of thirty degrees.

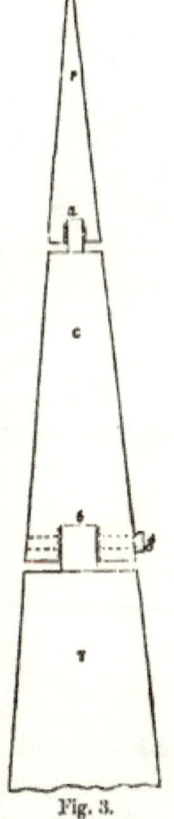

Fig. 3.

For the conductor of the 'paratonnerre' lengths of iron bars are principally used; formerly these were jointed together by means of a pyramidal bolt let into a notch of the same form, and connected by a simple iron pin. This method, however, was discovered to be very bad, as it failed to preserve the continuity of the conductor after it had been erected a little time. The following plan, as represented in fig. 4, is now used for the best work, as being more durable and affording a better contact. On each side of the bars to be joined, two flanges, about six inches long, and half the thickness of the bars, are filed out. A thin piece of carefully-prepared lead is then placed between them. The whole is then firmly fastened together by bolts at B and B and completely covered with pewter solder, and thus furnishes a solid, durable contact which possesses very small resistance.

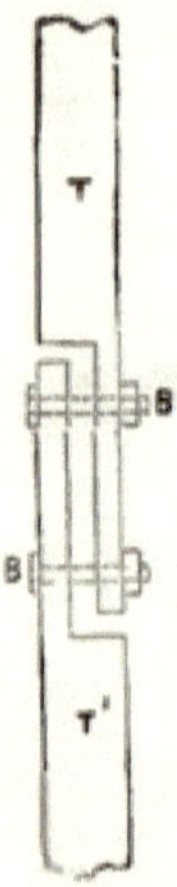

Formerly the conductors were, at regular intervals, rivetted to cramps let into the wall for the purpose of retaining the conductor in its place. As this plan left no room for the play of expansion and contraction caused by variations in the temperature, it was found that at times the conductor was very much strained and even bent by reason of this expansion and contraction. To avoid this evil an apparatus, which has been approved by the Paris Academy of Sciences, has been substituted for the cramps and rivets. This apparatus consists of a fork in which the conductor is held fast by a pin (see fig. 5). Being able to move backwards and forwards in

Fig. 4.

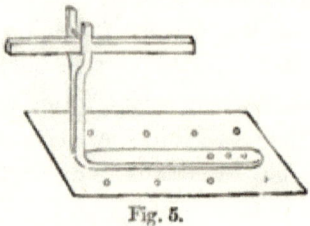

Fig. 5.

the fork with great facility, the conductor is thereby permitted to expand or contract under the influence of temperature without threatening its supports with destruction.

The question however arises, upon what part of the paratonnerre ought the effect of such contraction or expansion to be borne? The Paris Academy of Sciences has sanctioned and recommended the use of a compensator, which is designed to bear this strain. This compensator, which is now much used in France, may be seen in fig. 6.

It is composed of an elastic plate F, made of well-annealed red copper, three-quarters of an inch wide, at least twenty-eight inches long, and about a quarter of an inch thick. The two extremities of this plate are firmly fastened to the two ends of two lengths of the conductor by the bolts and counterpieces B B', and afterwards covered with a thick coating of pewter solder. When, in consequence of the heat, the conductor expands, the curve of the copper plate F will become greater, and in cold weather it will become less. As a rule, a single apparatus is supposed to compensate for the effects produced by long straight lengths, and it is therefore hought sufficient to place one at each bend.

With the exception of the terminal rod, it is the rule in France to cover the whole of the paratonnerre with some coating in order to preserve it from contact with the air.

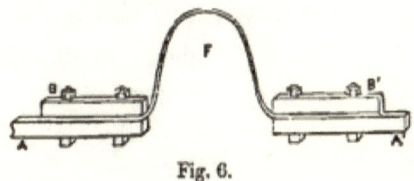

Fig. 6.

This is attained by covering it with either a strong coat of tar, or a painting of a metallic basis, such as zinc or tin filings.

In larger buildings what is termed a 'ridge-circuit' is often used. It consists of an unbroken metallic connection running along the ridges of the building to be protected, and connected with the conductors and terminal rods, and consequently with the subterraneous sheet of water which forms the common reservoir. It is made of lengths of square iron bars or rods having a thickness of about three quarters of an inch square, and fastened together by overlaying the ends, bolting them together with two bolts, and covering them well with solder in the manner shown in fig. 4. New branches are formed by T-shaped connections, the cross-piece of the T overlaying the original ridge-circuit, and the stem making the first length of the new branch. In some cases the ridge-circuit rests directly on the ridge of the roof; but in order to avoid injury during the repairs to the roof or in other ways, the plan adopted in good work is to raise it some distance above the ridge on supports at suitable distances, and thus prevent the possibility of damaging the joints and solderings.

The form and arrangement of these supports depend on the nature of the roof. Sometimes forked uprights are used —these allow for the expansion and contraction due to changes of temperature; in other cases simple cast-iron bearings, weighing from ten to twelve pounds each, are laid upon the ridge, their upper surfaces being grooved to receive the bars of the ridge-circuit.

All masses of metal used in the construction of the building are metallically connected with the paratonnerre. As a rule, this is done by pieces of iron about half an inch square, which are strongly soldered to the metal surfaces, and then connected with some part of the conductor or ridge-circuit.

Although in France, as elsewhere, all experts are agreed as to the prime importance of the disposition and arrangement of the *racine* or earth-end of the paratonnerre, a difference of opinion prevails as to the best means of insuring a good earth-contact, and many methods have been tried, all of them similar in principle, but differing somewhat in application. It is proposed to give here a brief outline of the best contrivances employed for this purpose.

One main object, in arranging the earth terminal of a lightning conductor, is to avoid the gradual destruction of the *racine* by the action of alternate dryness and moisture which, unless the iron is protected in some way, corrodes, and eventually eats it entirely through. There are several ways of remedying this evil. In France it is common to find used for this purpose a vertical spout of tarred, boucherised, or creosoted wood, rising a few inches above the soil. Some authorities recommend the simple plan of covering this part of the conductor with a strong coating of tar, others covering it with a wrapper of sheet lead, and this last method is probably the best. With regard to the extreme end of the conductor, the system approved of by the Paris Academy of Sciences is generally used in good work. This system is the use of a trough filled with broken charcoal, through which the conductor runs; charcoal preventing the too rapid oxidation of the iron. For charcoal, coke may be substituted. The trough (see fig. 7) is made either of wood, gutter tiles, or ordinary bricks without mortar, so as to allow the moisture of the soil to permeate through. It is preferable, even at the expense of lengthening the conductor, to carry it through the lowest and dampest plots of ground around the building.

To obtain a perfect contact between the end of the conductor and the earth, or common reservoir, the French use

Fig. 7.

several methods. One of the earliest ones was the multiplication of the iron bars attached to the end of the conductor, and inserting them for some distance into well-water. Theoretically this arrangement is good, but it has been found that the decay of these terminals by the action of rust was so rapid that, unless they were carefully watched and periodically repaired, they soon became insufficient, if not useless. In addition to this, it is the opinion of many French savants that a mere water contact is not enough, a soil that is always moist being in their judgment far preferable. The simplest plan adopted for attaining this end is that of inserting into the moist ground to a certain depth, regulated by the nature of the soil, one or several metallic branching stems, which are connected with the conductor. By another arrangement, invented by M. Callard, the conductor is terminated by a kind of galvanised iron grapnel, placed in a wicker basket filled with pieces of coke. Where the soil is dry, and moist ground cannot easily be got at, the harrow or grating shown in fig. 8 is often used. It is placed between two layers of horn embers, or charcoal, and sunk as deeply as it conveniently can be, the end of the conductor being carefully connected with it by soldering or by a quantity of melted zinc.

In towns, the water-pipes and gas-mains, possessing as they do, large metallic surfaces, are generally utilised for making the 'earth-contact.'

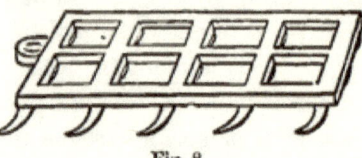

Fig. 8.

Sometimes, instead of iron bars, galvanised iron cables of about an inch in diameter are used for the conductors of paratonnerre, and occasionally red copper cables of half-an-inch only in diameter, but the use of these latter is uncommon.

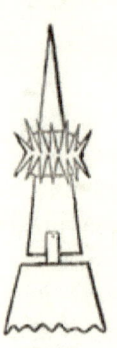

Fig. 9.

Fig. 9 exhibits a modification of the point of the terminal rod which is advocated by M. R. F. Michel. The arrangement is based on the principle that, on the approach of a tempest cloud, the more points there are, the greater will be the neutralising effect. M. Michel considers that when a terminal rod has only one point, it acts only in one direction; but if there is a large number of points branching in all directions, the preventive action is materially increased; he therefore proposes the use of this contrivance, which is carried out by having the ordinary conical trunk copper point on the top of the terminal rod melted down, and moulded so that it presents in its middle a circular swelling. Into this swelling arrows are fixed, inclined at each side of the horizontal plane to an angle of 45 degrees, as shown in the engraving. These arrows radiating in all directions are supposed to 'hasten the neutralisation of the electrified cloud;' and in the event of a discharge, the discharge, by dividing amongst them, will prevent their fusion.'

Before quitting the French system, mention should be made of a novel form of lightning conductor devised by M. Jarriant. This gentleman proceeds on the hypothesis that the most essential requisites of a lightning conductor are:— a terminal rod metallically homogeneous, which should rise to a good height; that it be sufficiently light to avoid damage to the roof, and yet be strong enough to resist the violence of the wind. To attain these requirements, M. Jarriant secures his conductor with three or four stays, which are firmly fixed to the roof and converge to the top of the terminal rod, to which is fastened the ordinary copper point, recommended in the 'Instruction' of the Academy of Sciences. Iron supports are placed at different heights in order to ensure the perfect solidity of the system. Galvanised iron is employed, and all the various stays and supports are metallically connected with each other. The angles of the irons are all acute, and placed so as to offer the least resistance to the wind. The advantages claimed for this method are that the upper part of the conductor bristles with spikes and aigrettes, which he considers a great advantage in regard to the preventive effect produced by the conductor; it allows of the conductor being raised much higher above the building; it presents a large surface to the electrified cloud; the joints are so arranged that they cannot be dislocated by the expansion and contraction caused by variations of temperature; and, lastly, it is affirmed that these conductors cost thirty per cent. less than those erected under the ordinary system.

America stands pre-eminent above other countries in the numerous and extraordinary schemes that have there been promulgated in regard to the protection of buildings from the effects of lightning, and probably no other nation has been so systematically victimised and swindled in this matter. The tramping 'lightning-rod men' of the United States are notorious for extortion and ignorance: they use all kinds of fantastic and peculiar shaped terminal rods and conductors, the main object being to make as great a show with as little metal as possible. Their work is almost entirely confined to the upper portion of the conductor, to the neglect of the most important part—the earth terminal. In consequence, the majority of the lightning conductors in America are untrustworthy; very often they are practically insulated from the 'common reservoir' or subterraneous water, and are therefore more often a source of danger than a protection. Unhappily, these peripatetic mechanics are by no means extinct, although increased knowledge is gradually driving them from the field.

In America, a strong point is made of utilising, as far as possible, all the existing natural conductors that are to be found upon a building, such as gutters, rain-pipes, and other metal surfaces. During a tempest, the opposite electricities of the earth and the air often select, by their inductive influence, a rain-pipe, gutter, metal roof, &c., for the passage of an electric discharge between them, and unless these metallic surfaces are connected with the earth, they are apt to be dangerous. But if they are properly connected together, and provided with a good earth-contact, they materially assist to diminish the intensity of a discharge.

In the case of a building with a roof of slate, wood, or other material of low conductivity, a conductor made of either bar iron or stranded cable is placed along the ridge and gable ends, and carefully connected with the gutters and rain-pipes; where the rain-pipes are less than three inches in diameter, the bar or cable conductor is often extended from the roof down the side of the building, and connected with the earth terminal. When this is done, the bar or cable conductor is placed between the rain-pipe and the wall of the building, or at any rate close to the rain-pipe, and connected with it by solder or bolts.

All metallic chimney caps, cornices and railings on the tops of buildings, as well as the water-pipes, gas-pipes, hot water-pipes, and other large or long pieces of metal, whether they occur inside or outside the building, are connected with each other by a conductor composed of light stranded wires, each about three-sixteenths of an inch in diameter; they are also connected with the main conductor at its nearest point. Where several adjacent buildings have each a metallic roof, these roofs are connected together by means of a horizontal conductor.

The terminal rod of the conductor generally projects about four feet above the chimney or other highest point of the building. It consists of a round iron rod seven-sixteenths of an inch in diameter, the lower extremity being hammered out for the purpose of fastening it to the conductor by soldering and screws or by bolts. A small building, not exceeding twenty-five feet in length or breadth, is generally fitted with either one terminal rod placed on the centre of the ridge of the roof, or with two rods, one at each end of the ridge, the latter method being the preferable one. In larger buildings terminal rods are placed at intervals of about twenty feet along the roof. The upper end of the rod is sometimes pointed, but not always, the argument being that although the ordinary end of a rod is blunt when used in connection with a Leyden jar, but that when applied to a thunder-cloud, which extends over thousands of acres, it becomes pointed, and bears the same proportion to a thunder-cloud as the sharp point of a needle does to the hand of a man. Occasionally the point is tipped with platinum, gold, silver, or pure copper, in order to prevent oxidation, but this is not considered essential, it being presumed that practically no amount of rust on the top would impair the efficacy of the terminal rod.

In the case of a building having a flag-staff upon it, a galvanised iron wire is fastened along it and projects about six inches above the top, the lower end of the wire being of course carefully soldered or otherwise connected with the main conductor.

Steeples and spires, in addition to the ordinary vertical conductor, are fitted with horizontal conductors placed around them at intervals of about twenty feet, and connected with the vertical conductor. This is to provide against the occasional discharges that take place in the centre of steeples, and which are caused by the deflection of the discharge in the air by the rain.

Chimneys and air shafts, from which heated air or smoke escapes, are fitted with metallic caps which are connected with the general conductor. In order to protect this metallic cap from the effects of the sulphurous fumes arising from the chimney, a terra-cotta cap is contrived to fit inside the metallic cap. An analogous method is adopted with regard to the ventilators of barns and ice-houses. If these buildings have the ordinary ventilators in the form of dormer windows upon the roof, an iron rod seven-sixteenths of an inch in diameter is placed vertically across, and above the centre of the opening of each ventilator, and connected with the conductor. Should the barn or ice-house have openings or doors through which warm vapour can escape, a conductor is fixed to the roof at the gable ends above the centre of each opening or door, and extended outwards about five feet, at an angle of forty-five degrees from the roof, so as to be in line with any ascending vapour, or any descending charge of electricity following the course of the vapour. All these auxiliary conductors and terminal rods are metallically connected with the main conductor of the building.

The conductors are simply fastened to the building by iron staples or by straps of sheet iron, pierced with two holes for nails or screws.

In America, as elsewhere, the earth-terminal is regarded as of prime importance, and in all properly constructed lightning conductors receives the greatest care and attention. In the first place, such metal pipes as lead from the building to the water-mains, gas-mains, and sewers are carefully connected with the principal lightning conductor, in order that they may act as auxiliary earth-terminals. For the principal earth-terminal many contrivances have been brought forward, but very few possess any originality, and many are positively useless. Some of the best are similar to those in use in England; among others, perhaps the best method is that of placing a cast or wrought-iron pipe of three inches inside diameter, and about ten feet long, vertically in thoroughly moist earth and carefully connecting the conductor, or conductors, with it. The chief objections to this plan are the occasional difficulty of getting a moist earth at all, and the possibility of earth that is generally moist getting dried up in hot weather. To obviate these risks, the following arrangement is used :—

In a pipe of wrought or cast iron, at least ten feet long, and having an inside diameter of two inches with a thickness of three-eighths of an inch, are made a number of longitudinal openings or perforations, about ten inches long and a quarter of an inch wide. These openings or perforations are made at intervals of ten inches, and are placed in one or two lines opposite to each other. If it is preferred, round holes of from half an inch to one inch in diameter, and about six inches distant from each other, may be substituted for the longitudinal openings. This perforated pipe is placed in an upright position in the earth, and is so situated that it receives at its top opening the waste or rain water flowing along a channel or drain constructed for that purpose. The water, after running into the top of the pipe, gradually percolates down, and passing through the perforations or openings into the earth around and underneath the pipe, moistens it to such an extent and at such a depth as to render it but little affected by the heat of the sun. The pipe is generally placed at some little distance from the building, so as to give a sufficient area of earth to be kept moistened and to prevent the walls of the building being affected by the damp.

Occasionally, the pipe is made triangular or square, and with perforated branches and other metallic conductors. It is also sometimes constructed with enlargements at the top or bottom, so as to hold more water. Probably, however, the simplest plan is the best, as—if the soil be suitable—a plain round wrought-iron pipe can be driven into the earth. If a cast-iron pipe is used, a hole of a convenient size is excavated for it. In this case, great care has to be taken that the earth is thoroughly well rammed down all round the pipe.

Another arrangement is to employ, instead of a cast or wrought-iron pipe, a number of round or flat-iron bars, fastened together at the top and bottom by rivetting to metal hoops in such a manner that intervals are left between each bar, through which the water can pass. Sometimes a solid pipe without the openings is used, but it is not found to be so satisfactory as the perforated pipe, because the latter allows a greater amount of water to pass through it into the soil, thereby furnishing a larger area of moist earth.

The French method of carrying the conductor to the bottom of a neighbouring well is frequently adopted where it is practicable, and the water of the well is not required for drinking or cooking purposes.

A few words may be added here on the method of protecting the large mineral oil tanks which are to be found in the United States. Many of these oil tanks are of very large capacity, some of them containing a million gallons of oil. They are generally constructed of thick iron plates rivetted together. The roofs are usually made of wood coated with tar, but in some cases iron is adopted. As a rule, several of the tanks are grouped together and connected with each other—and in some instances with distilleries—by means of subterraneous iron pipes.

One method of protecting these tanks is to erect around them, at a distance of some ten feet, wooden supports, on which are placed upright metallic conductors which overlook the tank, and are connected with each other near their tops by stout iron wires, thus forming a network of conductor which is supposed to intercept any discharge of electricity from a tempest-cloud, and prevent it from reaching the oil tank. This method, however, has failed in several notorious instances, and is not countenanced by the best authorities.

A better and less complex arrangement is now usually adopted by the best firms. The chief object of this arrangement is to prevent the temperature of the oil tank, and of the atmosphere above and around it, being raised by means of an electric discharge. This is accomplished by using large conductors, which are carried some distance above the oil tank. These conductors, of which there should be at least four, are formed of flat iron bars about one and a half inches wide and half an inch thick; they are securely fastened to the sides of the tank at equal distances from each other, and metallically connected with it. About thirty feet above the roof of the tank they meet, and are carefully and substantially joined together, and supported, if necessary, by a wooden post extending from the centre of the roof of the oil tank.

The earth terminals, of which there must be one to every two conductors, consist of perforated iron pipes, as before described, three inches in diameter and fifteen feet long. They are sunk into thoroughly moist earth, and metallically connected with the lower part of the tank. These perforated pipes are so arranged that they catch the rain water from the roof of the tank; by this means the surrounding earth is kept moist. It may be mentioned that by utilising the tank as a portion of the system of conductors, the electric discharge is distributed and much weakened.

CHAPTER XIII.

NEWALL'S SYSTEM OF PROTECTING BUILDINGS.

IN its essence there cannot be anything more elementary than the theory of protection against lightning. It is simply to lay a metallic line from the top of a building, or other object to be protected, into moist ground, so as to make a path for the electric force, along which, not finding impediments, it will travel freely, without causing the least damage. But, like many other simple theories, their practical execution is not without perplexities. The first of these, in regard to conductors, arises from the existence of more or less considerable quantities of metals, to be found in almost every building which requires protection against lightning. As the use of metals, especially iron, in the construction of dwellings, both exterior and interior, is rapidly extending, this becomes a very important consideration in planning the design of lightning conductors. Of equal moment is a second point—that of the existence of water or great moisture under the buildings, or part of them. This must decide invariably the direction of the conductor towards the earth, and its depth underground. There are many minor matters to be taken into account, but these two may be laid down as the chief questions to be kept in view in settling the best mode of application of any conductors under given circumstances. It happens often enough that a proper solution as to what is best is not a little difficult. Still, it can always be arrived at by careful study, which must, however, be aided by experience.

Keeping always in view the fact that there is nothing whatever that may be called 'erratic' in the manifestations of the electric force, but that it acts under a ruling principle as strict as that governing the law of gravity, the first point in designing the protection of any building will be to clearly ascertain what path the lightning will take on its course from the clouds to the earth. It is absolutely certain that the electric force will make its way through materials, termed good conductors, which allow it free passage, and avoid those of the opposite class, or bad conductors, the character of every substance on earth being well known as regards these qualifications, although it would not be easy to draw sharp lines of demarcation, all conductivity being relative and not absolute. Looked at in this way, the fundamental one in the application of lightning conductors, the simplest object for protection will be a pyramid of stone, such as the Egyptian obelisk, popularly called 'Cleopatra's Needle,' erected on the Thames Embankment in the summer of 1878. Stone being a bad conducting material, all that is necessary to protect it against lightning, provided there is no metal whatever near it, is to run a thin strip or rope of copper from the summit to the base, and down into moist earth. Although fragile, the strip of copper, if uninterrupted and rooted in moisture, will in this case form an absolute protection. The question assumes another aspect if, instead of a stone pyramid, a tall factory chimney, not very dissimilar in outward form, is given as an object for protection. Here there enters another element. A tall pile of bricks is as bad a conductor of electricity as a solid mass of stone, but the mass of bricks constituting a factory chimney is hollow, and the cavity being filled with smoke and mineral fumes, which are more or less good conductors of the electric force, the artificial path laid for the free passage of lightning has to surpass in acceptability the natural one. In other words, the copper rod laid alongside the factory chimney, to secure it against damage from lightning, must be considerably thicker than the one which will protect the simple stone pyramid. It is this principle which has to be followed all through in the application of conductors. They must form, in one word, the best path which can possibly be made for the electric force.

The system employed by Mr. R. S. Newall, F.R.S., for the construction and erection of lightning conductors is probably the most complete—and certainly the most representative—of the various methods in vogue in England. The special study Mr. Newall has made of the subject in all its bearings, both theoretical and practical, added to the fact of his possessing at his extensive cable works at Gateshead such exceptional facilities for the production of copper ropes and bands composed of the purest metal, render him one of the first authorities on all matters connected with the application of lightning conductors to buildings. In describing, therefore, the English method, reference will chiefly be made to this gentleman's apparatus and inventions.

The function of a lightning conductor is twofold. In the first instance, it operates as a medium by which explosions of lightning, or, to speak more accurately, disruptive discharges of electricity, are led to the earth freely, and without the risk of their acting with mechanical force, as they invariably do when compelled to pass on their way to the earth through so-called non-conductors, that is to say, bodies possessed of low conductivity, such as the atmosphere, wood, stone, &c. In the second instance, the conductor acts as a means whereby the accumulation of electricity existing in the atmosphere is quietly drawn off and carried noiselessly into the earth, and dissipated in the subterraneous sheet of water beneath it. Now this accumulation of electricity, always greatly intensified during a thunderstorm, invariably seeks the easiest road to earth; this road is technically called 'the line of least resistance.' This line of least resistance is influenced by various circumstances; the resistance of any line may be lessened by the presence of streams of warm vapour or rarefied air such as would come from chimneys, from barns or stacks containing new hay;

by a column of smoke, or by the presence of tall trees moist from rain. It is not always easy to find the reason why the lightning takes any particular path, but one thing is certain, that is, it acts under certain fixed principles, and does not take any particular route by chance, but always because it is the line of least resistance. What the lightning conductor really does is to prevent the possibility of an electric discharge within a certain district, for instance, in the interior of a house or other building.

From the above remarks, it will easily be seen that lightning conductors should be made of materials possessing the highest possible power of conductivity, and be large enough to carry off the heaviest electric discharge that is ever likely to fall upon them. The various metals being by far the best conductors of electricity, it follows that the lightning conductor must be constructed of metal of some kind. But even metals differ to a great extent in their conducting powers, as has been shown in a previous chapter. There are, however, only two metals which are practically available for use as lightning conductors, namely, iron and copper, and after repeated experiments Mr. R. S. Newall has arrived at the conclusion that a conductor made of copper of adequate size is the best—and, in the end, the cheapest—means of protecting buildings from the effects of lightning. The relative conductivity of iron and pure copper being as six to one, it follows that if a copper cable or bar of a given size be sufficient, an iron cable or bar ought to weigh six times as much per lineal foot in order to be equally safe. It may be added, that while copper is more expensive, weight for weight, than iron, it is not so liable to oxidise; nor, on account of its higher conducting power, is it so easily fused. The comparative smallness of its mass renders it far more manageable than iron, and does not interfere with the architectural features of the building on which it is used. On the contrary, it is readily adapted to curves and angles.

It may therefore be taken for granted that, almost without

exception, pure copper is the best material that can be used in the construction of lightning conductors.

The size of the terminal rod or point used in Mr. Newall's method varies in length and diameter according to the

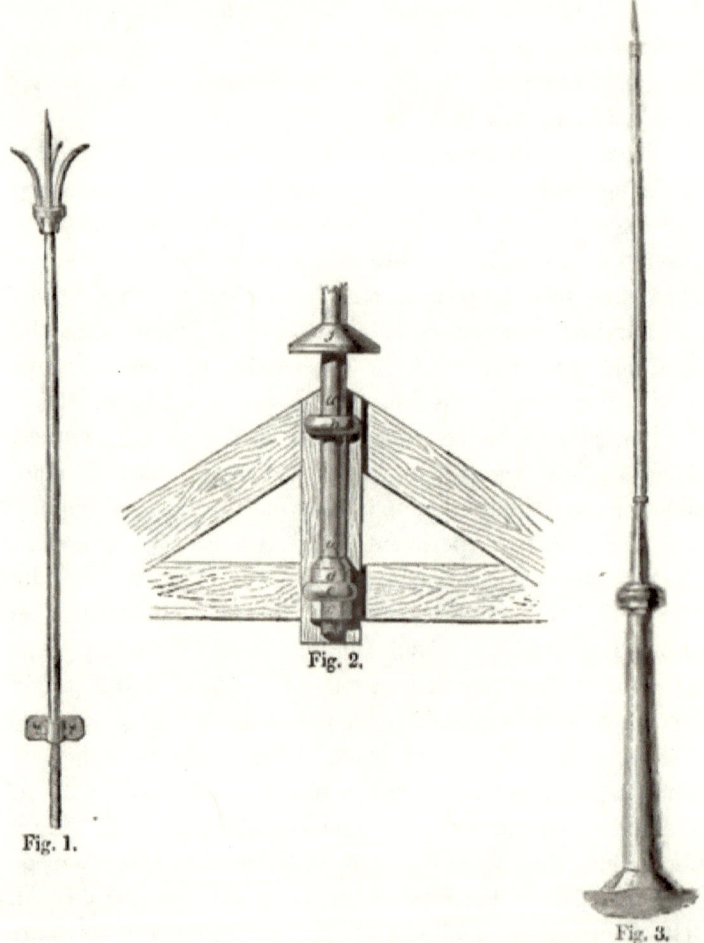

Fig. 1. Fig. 2. Fig. 3.

extent and height of the building to be protected. As a rule, they are from three to five feet in length, and from five-eighths to three-quarters of an inch in diameter; at the upper end they branch out as shown in fig. 1.

In conjunction with this terminal rod a short description of the 'Auffangstange,' or 'reception rod' of the Germans, may be given. This 'reception rod' (see fig. 3) is made of iron, and varies in length from ten to thirty feet. It consists of two parts, the higher part, which measures two-thirds of the whole length, is fastened by a flange to the lower part of the rod. In fixing this German 'reception rod,' its height and weight have to be taken into consideration. It is generally made fast by two strong staples, b and c, as shown in fig. 2, which pass through the king post of the roof and are fastened behind by screw-nuts. The part marked d rests in the lower ring c so that it cannot sink, and the extreme end passes through this ring c and is screwed tightly to the nut e; f is a cap to prevent the rain getting into the roof.

It is much to be regretted that not only professors and amateurs studying the manifestations of the electric force, but even learned societies, such as the French 'Académie des Sciences,' should have spread so many imaginative theories about this 'reception rod.' At the bottom of all was the fancy, not often declared, but still visible in its expression, of the metallic conductor possessing some occult power of *attracting* lightning. In France, as well as in Germany and Italy, there existed for a long time, and to some extent still exists, quite a mania for erecting huge rods, such as that shown in the engraving (see fig. 3), on the top of buildings, the general belief being that the more high-towering the greater would be the 'area of protection.' A little common sense, brought to the aid of fanciful imaginings, should have taught the supporters of this 'area-of-protection' theory that it was absolutely untenable. The electric force, seeking its nearest path to the earth, could not be expected to diverge from it through the action of a rod raised somewhat higher than the surrounding building; and the proper method clearly was to bring the metal everywhere as near to any possible emanation of the force, whether lateral or vertical, as could be done. Besides being really of no use, except in rare instances, such as the neighbourhood of high trees, these tall rods formerly employed, and still frequently seen on the roofs of buildings, had the detriment of being unsightly, while at times they were positively dangerous. Instances occurred in which a high wind threw them down from their elevated position into the road below, on the heads of passers-by. Thus two persons were killed in Paris in the summer of 1830 by the fall of a gigantic ' tige ' from the steeple of the church of St. Gervais. Either at the same moment, or immediately before, a stroke of lightning fell upon the church in its lower part, away from the conductor, making a hole in one of the walls, and then escaping, without doing further damage, by some iron water-pipes running underground. The conductor in this case had been constructed on the model approved by the 'Académie des Sciences,' but the accident conclusively showed that there was no trust to be placed in any mere theoretical calculations as to the extent of the ' area of protection.'

A noteworthy example of the fallacy of the 'area-of-protection' theory is to be found in the case of the explosion at the powder magazine at the Victoria Colliery, Burnt-cliffe, Yorkshire, which was struck by lightning and destroyed on August 6, 1878. The instance is also instructive as showing how important it is that copper conductors should possess the highest possible conductivity—i.e. be made of the best and purest copper.

The magazine was an oblong building of brick, nine feet long, five feet wide, and six feet high (internal dimensions), and it had a uniform thickness of three bricks. At one end was a heavy iron door, and at the other a lightning conductor, consisting of a copper-wire rope seven-sixteenths of an inch in diameter. The point of the terminal rod was about thirteen feet above the top of the building, and a similar length was carried into the ground and terminated in clayey soil. The conductor was fixed to a pole distant about two inches from the end of the building opposite to that in which the iron door was fixed. *It was not connected with the iron door in any way.* At the time of the explosion the magazine contained about 2,000 pounds of gunpowder.

Major Majendie, H.M.'s Chief Inspector of Explosives, in his official report ascribed the accident to the fact of the iron door being unconnected with the lightning conductor, and in doing this he was doubtless right, but only to a limited extent. The author of this work visited the colliery shortly after the explosion, and found that the conductor— the weight of which was about one pound per yard—had been fastened to the pole, which was about twenty-one feet high, by two glass insulators, and that the conductor was not connected with the building. On testing the copper rope which formed the conductor, its conductivity was found to be only 39·2 instead of 93 or 94 per cent. The conductor, therefore, was but little better than if it had been made of iron, and, even supposing it had been made of good copper, it was of too small a size. It should have been of double the weight, and *not* insulated from the pole. In order to be thoroughly efficient it ought to have been brought down the pole, carried through under the roof, down the iron doorpost, and so into the ground.

According to the French theory, that the 'area of protection' afforded by a lightning conductor is the space contained within the circular area of a radius double the height of the conductor, the magazine was thoroughly secured, for the conductor was twenty-one feet high, and the building only nine feet long, five feet broad, and six feet high. This case, however, with many others, entirely controverts this theory, and shows very forcibly the fallacy of an argument that at one time was accepted almost as an axiom.

One other case of more recent date may be instanced. At Cromer, in Norfolk, the church—a fine perpendicular building of flint and freestone, having a tower 159 feet high —was damaged by lightning in August 1879. During a thunderstorm the lightning struck one of the pinnacles with considerable force, although on another pinnacle, only twenty-seven feet six inches distant, a good copper conductor, having a diameter of five-eighths of an inch, was fixed. On testing the conductor by means of a galvanometer, it and the earth connection were found to be in thoroughly good order. After what has been said, comment on this last example is needless.

The general disposition and adjustment of a lightning conductor demands the greatest care and consideration. No hard and fast rules can be laid down, for each individual case must be studied and elaborated by itself, especially in the instance of large structures, where much depends upon

Fig. 4.

style, outline, and other details. The main point is that *every* part of the building shall be placed beyond the possibility of being damaged by a disruptive discharge of electricity.

It has been stated previously that the lightning invariably follows the line of least resistance, and that this line may be influenced by the presence of streams of warm vapour, columns of smoke, &c., which, escaping into the air, furnish a ready path for the electric discharge. Consequently it sometimes happens that a building or barn may be struck although it be provided with a lightning conductor. In order to explain this it must be borne in mind that the line of least resistance is not always the shortest line mathematically. The accompanying illustration (fig. 4) is an example to the point. It represents a barn furnished with a lightning conductor and filled with new-made hay, which is a better conductor of electricity than the material of which the barn is constructed. This hay is giving off the stream of warm vapour which is pouring out of the opening at the end, and forms an invisible band of conducting matter between the thunder-cloud and the barn, as marked out in the engraving by the dotted lines, the direction of the wind being shown by the arrow and the trees. Under these circumstances the discharge of lightning would naturally follow the path between c and d in preference to the shorter route between a and b, because the former is the line of least resistance between the cloud and the earth. Thus the barn—although furnished with a conductor in good condition—would most likely be set on fire, or otherwise damaged. The same deflection of the lightning-stroke might be caused by a column of smoke, or by the fact of one portion of the building being moistened by the rain and the other kept dry; an occurrence that might easily happen when a strong wind is blowing during a storm.

In order to ensure complete protection, the conductor on the barn should have been carried along the ridge and down the edges of the roof at each gable. By this means the stroke of lightning would have been intercepted.

The engraving on the next page shows a design for the protection of a large detached mansion by means of a multiplication of short points or terminal rods fixed on all the prominent features of the building. The conductor is carried along the ridges in every direction, and down the edges of the roof at each gable. Generally it is sufficient to have two descending conductors, but occasionally the conformation of the building or the nature of the ground renders necessary the use of even more. It is imperative, for obvious reasons, that the descending portion of the lightning conductor shall be carried from the roof to the ground by the shortest possible route, and placed in perfect electrical contact with the earth in the manner to be indicated in a succeeding chapter.

Fig. 5.

The projecting points of the conductor are drawn in fig. 5 larger than they need be, in order to show them more clearly, distinguishing them from the rest of the building. The same has been done with the copper rod, running from the roof to the ground and thence into the earth. In reality a conductor may be made perfectly safe, and yet all but invisible to the naked eye. For private houses and buildings, a rope made of copper ought to be at least five-eighths of an inch in diameter, for a copper rod of half an inch in diameter has never been known to be fused. For chimneys of manufactories, where gases are liable to corrode the rope, it had better be a little thicker. Such copper ropes as those manufactured by Messrs. R. S. Newall and Co., five-eighths of an inch in diameter, weighing two-thirds of a pound per foot, and having a conductivity of 93 per cent., have never been known to fail in protecting even

Fig. 6.

the largest buildings. It is supposed by some writers that the value of the conductor is in proportion to the amount of surface of metal exposed. This, however, is a mistake, for the conductivity depends on the weight per foot of metal used, the purity in both being equal. Wire-rope is used simply because it is so pliable that it is easily handled, and can be made of any length required without joints.

In fig. 6 is given an illustration of a small detached house, in which the arrangement of the lightning conductor is indicated by the dark lines. The method followed is exactly the same in principle as that employed for the mansion just described. A terminal rod is placed upon each chimney. These terminal rods are connected with each other by a copper-rope conductor which is carried along the ridges and gables of the roof, thus constituting a similar arrangement to the French 'ridge-circuit' (*circuit des faîtes*), with the additional advantage of being far lighter and more sightly. The copper conductor descends to the earth down the angle formed by the projecting entrance to the house. By this means every corner of the building is protected; an important matter in all detached buildings, and especially when they happen to stand among trees. The preference of the electric force for trees as its path to the earth in the absence of metal or other bodies of higher conductivity than trees, has probably no other ground than their being full of moisture; still this is a disputed question.

Fig. 7 exhibits a slightly different method of arranging the lightning conductor. In this case the ridges of the roof are surmounted by ornamental iron-work, instead of the usual terra-cotta, or earthenware, tiles. This iron-work is utilised and carefully connected with the conductor. The chimneys, in place of being fitted with terminal rods, are provided with cast-iron caps—as shown in the engraving—to which the conductor is attached. The conductor, after descending to the ridge, is led along it and down the edges of every gable, and is finally carried down to the ground and connected with the earth in the usual manner. It is of course absolutely necessary that all masses of metal, such as gutters, waterspouts, rain-pipes, &c., should be brought into connection with each other and with the conductor, in order that the house may constitute one electrically homogeneous body.

It was for a long time held that the protection of churches against lightning offered special difficulties. This arose mainly from the constant reports of churches being struck, often when they were believed to be protected, whereas the accidents arose from the conductor not being properly fitted. It is even now too often forgotten that all so-called 'conductors' of the electric force are only so in relation to 'non-conductors,' and that, strictly speaking, all things on earth are to some extent conductors and to some extent non-conductors. This being kept clearly in view, there is no more difficulty in protecting the largest cathedral against lightning in the most efficient manner than in similarly guarding the smallest cottage.

A case in point occurred in May 1879. The steeple of

Fig. 7.

the church at Laughton-en-le-Morthen was struck by lightning and damaged, the lightning conductor being thrown down and broken into two pieces. A correspondence on the subject ensued in the *Times*, and Mr. R. S. Newall had the remains of the conductor examined, with the following result:

'The spire is 175 feet in height, and it had attached to it a thin tube, made of corrugated copper, about seven-eighths of an inch in external diameter and five-eighths internal.

The copper is about one-thirty-second of an inch in thickness, and it weighs about one and a quarter pound per yard. It is made in short lengths, joined together by screws and coupling pieces, but there is no metallic contact whatever between the pieces, which are much corroded.

'The conductor appeared to be fastened to the vane. It was not in contact with the building, which it ought to have been, but it was kept at a distance of about two-and-a-half inches from it by twenty-one insulators. The earth contact was obtained by bending the tube and burying it in the ground at a depth of from six inches to eighteen inches, the soil being dry loose rubbish; the length of the earth end was only three feet, with two short pieces of about a foot in length each tied to the tube by thin wires, thus forming altogether a most inefficient conductor. It was placed in a corner formed by a double stone buttress, which came between the conductor and a lead-covered roof attached to the spire, the distance between the conductor and the lead roof being about six feet six inches.

'The lightning appears to have come down the conductor a certain distance, and, finding the road to earth bad, it passed through the buttress, dislodging about two cart-loads of stone, and then came down the cast-iron down pipes leading from the lead-covered roof and so to earth.'

Mr. Newall, in writing to the *Times*, goes on to say :—

'Now if the conductor had been made of copper-wire rope, weighing about two pounds per yard, and fixed in contact with the spire, without insulators and with a proper earth contact, no damage whatever would have been sustained by the building; and if the conductor had been tested periodically by an expert he would have shown whether the conductor was good or useless. This examination ought to be insisted on, as the earth connection is often wilfully destroyed; but I have never in all my experience known a building which had a conductor properly fixed to suffer damage from lightning.'

What is really required is to make a lightning conductor

of sufficient calibre to carry down the electric discharge, however great it may be, from the summit of the building into the earth, and that the earth contact should be above suspicion and thoroughly good in all seasons.

Fig. 8.

Fig. 8 shows a plain and simple design for protecting an ordinary church. The conductor in the case of churches and all other high or extensive buildings ought invariably to be made of copper rope, other metals of less conductivity, such as iron, being inadmissible, since

Fig. 9.

their employment would necessitate the use of ponderous masses of metal, which would be not only unsightly, but extremely heavy, and difficult to manipulate successfully.

Fig. 10.

In the accompanying engraving (fig. 9) lent by the Society for Promoting Christian Knowledge, is shown a somewhat more complex structure and the method of arranging the conductors thereon. In this case there is a conductor attached to each spire, leading to and connected with the metal-work of the roof and gutters. On the gable c, and the transept gables $d\,e$, there are fixed three conductors which unite in the centre of the roof, from which they are carried down to the gutters. The same arrangement is followed for the smaller gables $f\,g\,h$. The water-pipes and gutters being connected with the conductors, these latter are carried down the side to the earth. It need scarcely be explained how important it is that all metal ornaments on the ridges of churches, as well as other buildings, should either be connected with the general conductors or, in the case of extensive buildings, with a conductor that is carried straight to the earth, as shown in fig. 10. In the case of the finials so often found on Gothic structures, it is necessary to splice the conductor round the bottom of

Fig. 11.   Fig. 12.

the finial, as shown in fig. 11. If, instead of placing terra-cotta tiles along the ridges, a cresting of fancy iron-work is fixed there, the expense of running a conductor along the ridges will be saved.

The various methods of fixing weather-cocks on to the terminal rod are fully explained in another chapter. Fig. 12 shows the best arrangement for connecting the conductor to the terminal rod on a church spire. The copper rope which forms the conductor is spliced round the terminal rod at the bottom of the finial, and as an additional security round the base of the vane rod, which in this instance also serves as the terminal rod of the lightning conductor.

There has been much controversy as to whether it is better to carry the conductor from the roof to the ground inside a building than outside the walls. As a matter of fact, it is a question of very small importance which way the conductor is carried, so long as it arrives at the ground by the shortest possible route. Benjamin Franklin, to judge from many expressions in his works, seems to have been decidedly in favour of the inside plan, which was adopted almost universally in France and on the Continent in general on the first introduction of lightning conductors. But the method was soon abandoned, owing partly to a witty saying of Voltaire, constantly quoted to this day. Speaking of the death of the unfortunate Professor Richman, of St. Petersburg, killed while experimenting with electric discharges from the clouds, Voltaire remarked, 'There are some great lords whom one should only approach with extreme precaution: lightning is such a one.' A mere jocular exclamation, it would have had no great force except in France, where a *bon mot* may cause the fall of a king and the dethronement of a dynasty. In regard to Voltaire's pleasantry about not approaching too close to lightning, it really had in great part the effect of preventing conductors to be laid inside the houses. Even such calm philosophers and men of science as Professor Arago quote Voltaire with approval. 'I feel inclined,' he remarks in his 'Meteorological Essays,' ' to admit that the illustrious author (Voltaire) may be right, when I remember a case that occurred in the United States.' The case relied upon, a very curious one, was as follows, in Arago's own words.

'Lightning,' Professor Arago tells his story, 'having struck a rather thick rod erected on a Mr. Raven's house, in Carolina, United States, afterwards ran along a wire carried down the outside of the house to connect the rod on the roof with an iron bar stuck in the ground. The lightning in its descent melted all the part of the wire extending from the roof to the ground-storey, without injuring in the least the wall down which the wire was carried. But at a point intermediate between the ceiling and the floor of the lower storey things were changed: from thence to the ground the wire was not melted, and at the spot where the fusion ceased the lightning altered its course altogether, and, striking off at right angles, made a rather large hole in the wall and entered the kitchen. The cause of this singular divergence was readily perceived, when it was remarked that the hole in the wall was precisely on a level with the upper part of the barrel of a gun which had been left standing on the floor leaning against the wall. The gun barrel was uninjured, but the trigger was broken, and a little further on some damage was done in the fire-place.' Commenting upon this case, Professor Arago goes on: 'Here the lightning went off horizontally through the wall, in order to strike a fowling-piece standing upright in the kitchen. How much injury might not have resulted from this lateral movement, if the lightning had not had to traverse a thick wall?' Consequently, he argues, Voltaire is right in his jocular-oracular declaration about the perils of indoor lightning conductors, in their being 'great lords' dangerous to approach.

It is really difficult to understand how a man like Professor Arago could be misled into such false reasoning as this about an accident which, in itself, was of the simplest, and of the very easiest explanation. That the stroke of lightning falling upon Mr. Raven's house, in Carolina, should have melted the wire of the conductor points to one cause, and to one only, namely, that there was no proper earth connection. Had it existed, the wire, although thin, could not possibly have been 'melted all the part extending from the roof to the ground-storey,' nor could the electric force have left its appointed path to seek a better one through a wall, and, still more astounding, ' striking off at right angles.' It is abundantly clear that such cases, and others to the same effect, brought against the fixing of lightning-conductors inside the walls of buildings, prove absolutely nothing. What is beyond controversy is, that a good conductor, in proper condition, is absolutely harmless to surrounding objects, including human beings. A man, even with a 'fowling-piece' in his hands, might lean full length against half-an-inch copper rod carrying off a heavy stroke of lightning into 'good earth' without so much as becoming aware of the passing of the electric discharge. If certainly a 'grand seigneur,' as Voltaire remarks, the electric force has this in common with some of the greatest of men, of not wasting its time, but following a clear aim.

A very common, and, it may be added, a very mischievous opinion is prevalent, that lightning conductors should be carefully insulated from the buildings to which they are attached, and consequently many conductors are made to pass through insulators of glass and other materials of low conductivity. This practice of separating the building from the lightning conductor is not only utterly useless but positively dangerous. It is not unusually thought that by insulating the conductor the electric discharge will be prevented from entering the building. Such an idea is *ipso facto* absurd, for it is preposterous to suppose that a flash of lightning which can travel through thousands of feet of air—itself a very bad conductor of electricity—and then shatter to pieces the most compact bodies, would be stopped in its course by means of a few inches of glass, or a few feet of air. It may therefore be confidently asserted that no insulator can possibly be made that would be capable of preventing the electric discharge leaving the lightning conductor provided it could find an easier path leading to the earth. Mr. Phin, in his work on 'Lightning-Rods' says very pertinently:—But not only are insulators worthless—they are positively dangerous if the principle upon which they are adopted is fully carried out, which, however, is but rarely done. A little consideration will show this. Thus, if a house be furnished with a carefully-insulated lightning-rod, and should also have any large

NEWALL'S SYSTEM OF PROTECTING BUILDINGS. 161

surface of metal, such as a tin roof, an extensive system of gutters, or such like, connected with it, it is easy to see that the house must resemble a large Leyden jar, of which the tin roof, or other mass of metal, constitutes one coating, and the lightning-rod and the earth constitute the other, while the insulators and the dry material of the house represent the glass of the jar. If both the outside and the inside of this jar (the tin roof and the earth) had been connected together, it would have been impossible to have brought one coating into a condition opposite to that of the other. But the rod being carefully insulated from the roof, it is obvious that the inductive action of the cloud will bring the roof and the earth into opposite conditions; and if a man were to form the path of least resistance between them, the discharge would take place through his body, and he would probably be destroyed. It is obvious, then, in the first place, that lightning-rods should be connected with all large masses of metal which may exist in or upon the house, such as metallic roofs, tin or iron gutters, or pipes, iron rail-

Fig. 13.

Fig. 14.

M

ings, &c. In the second place, the rod should be attached to the house in the neatest and least obtrusive manner possible.'

It is indeed desirable for various reasons that the copper rope or band forming the lightning conductor should be affixed to the building in the neatest and least obtrusive manner possible. The conductor may be fastened by means of ordinary metal staples made of stout copper wire. A better method however is indicated in figs. 13 and 14, one showing the rope conductor formed of forty-nine wires, usually employed by Messrs. R. S. Newall and Co. for the protection of ordinary houses and buildings, and the other the copper band used by them for the same purpose. This fastening is simply a strap of copper bent to the required shape and pierced with two holes, by means of which it is fixed to any building by copper nails or screws. This method possesses several advantages; it is very sightly and neat, it can be easily applied without injury to any building, and as it allows the conductor a certain freedom of movement, it readily permits the contraction and expansion caused by the variations of temperature. The band conductor shown here is one inch wide by one-eighth of an inch thick, and weighs ·44 pound per foot. The rope conductor, although it appears less, has more metal in it; it measures five-eighths of an inch in diameter, and weighs ·67 pound. Fig. 15 shows a different mode of attaching the lightning conductor. It is generally used for the heavier ropes.

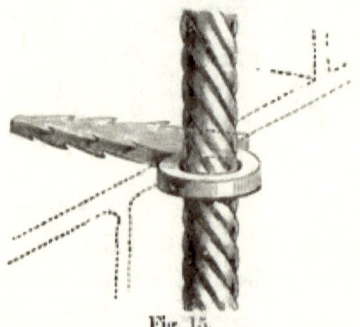
Fig. 15.

Fig. 16 exhibits an apparatus called a 'tightening screw.' It is used for making the conductor taut when it gets loose from any cause. The diagram explains itself, so there is no necessity for describing it.

The tall chimney shafts of factories and similar buildings, from which smoke or rarefied air escapes, are peculiarly liable to be struck by lightning. This is principally due to the current of smoke or warmed air forming, with the soot in the chimney, a medium conductor leading to the iron-work of the furnace or stove beneath, but ending there—a result that must be carefully avoided; for although a conductor that leads past any object is a protection (provided

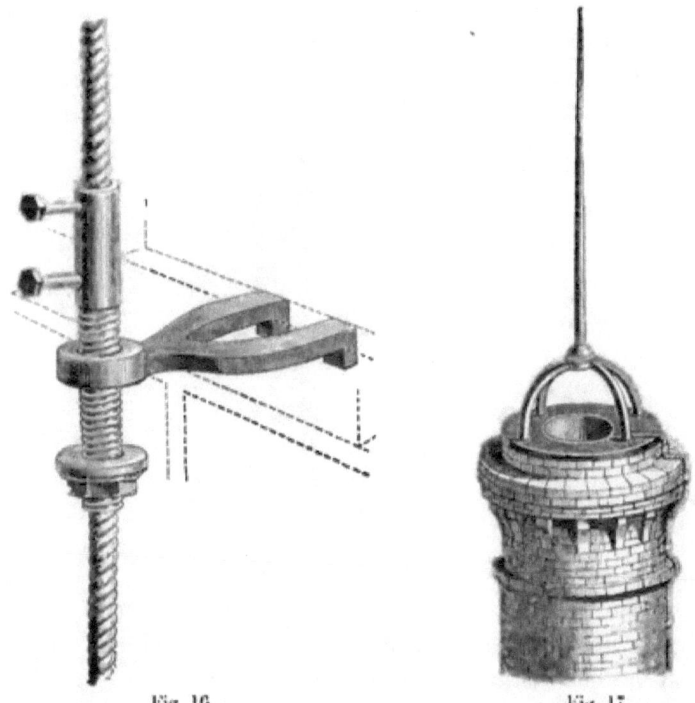

Fig. 16.   Fig. 17.

always that it has a good earth connection), a conductor that leads to an object, and ends in that object, is a distinct danger. It is therefore necessary to offer to the electric discharge a better conductor, able to intercept it and convey it safely to earth on the outside of the shaft.

The mode by which this is generally accomplished in England is by fixing a copper terminal rod (four or five feet long), on to the side of the top of the chimney shaft. This method is open to one serious objection: if the wind should happen to blow the stream of smoke or heated vapour in a direction opposite to the terminal rod, the electric discharge might go down the chimney shaft and effect considerable damage. By far the best plan is that shown in fig. 17. It consists simply of an iron or copper cap, to the centre of which is attached the terminal rod. This latter, however, is by no means essential, and may be said to be merely placed on the top for ornament. A structure of such small circumference really wants no terminal rod, the most important thing being to provide a copper rope or band conductor of sufficient size to carry

Fig. 18.

any electric discharge in safety to the ground. It will conduce greatly to the strength and stability of such a conductor if it be built up together with the chimney shaft, and fastened into the brickwork by clamps on the plan shown in fig. 18. A conductor of this kind should be made of copper rope or band of much greater calibre and weight than that used for ordinary buildings. That made of seven solid wires twisted together (see fig. 19) being the best.

A theory propounded some years ago by the late Prof. Clerk Maxwell, F.R.S., one of the most eminent physicists in Europe, deserves some notice here, perhaps more from its ingenuity than its practical accuracy. On investigation, it proves to be a revival of an old presumption that it is possible to protect a powder magazine or other building from the effects of lightning by having its roof, walls, and ground floor surrounded with a covering of sheet metal, or a network of lightning conductors, and disconnecting the said covering or network from the earth, or even insulating it by means of a layer of asphalt or some similar substance. Prof. Clerk Maxwell argues that the presence of a lightning conductor induces a larger number of electric discharges in its immediate neighbourhood than would occur provided no conductor was present, although at the same time these discharges are rendered less intense and smaller by reason of the existence of the conductor. Therefore, it is possible that fewer discharges take place in the area just outside the radius of the conductor. Reasoning from this, Prof. Clerk Maxwell considers that an ordinary lightning conductor tends rather to mitigate the accumulation of electricity in the clouds than to protect the building on which it is placed.

Fig. 19.

He says: 'What we really wish to prevent is the possibility of an electric discharge taking place within a certain region—say, in the inside of a gunpowder manufactory. If this is clearly laid down as our object, the method of securing it is equally clear.

'An electric discharge cannot occur between two bodies unless the difference of their potentials (i.e. their electrical conditions) is sufficiently great, compared with the distance between them. If, therefore, we can keep the potentials of all bodies within a certain region equal, or nearly equal, no discharge will take place between them. We may secure this by connecting all these bodies by means of good conductors, such as copper wire ropes, but it is not necessary to do so, for it may be shown by experiment that if every part of the surface surrounding a certain region is at the same potential, every point within that region must be at the same potential, provided no charged body is placed within the region.

'It would therefore be sufficient to surround our powder-mill with a conducting material, to sheath its roof, walls, and ground-floor with thick sheet-copper, and then no electrical effect could occur within it on account of any thunderstorm outside. There would be no need of any earth connection. We might even place a layer of asphalt between the copper floor and the ground, so as to insulate the building. If the mill were then struck with lightning, it would remain charged for some time, and a person standing on the ground outside and touching the wall might receive a shock, but no electrical effect would be perceived inside, even on the most delicate electrometer. The potential of everything inside with respect to the earth would be suddenly raised or lowered as the case might be; but electric potential is not a physical condition, but only a mathematical conception, so that no physical effect would be perceived.

'It is therefore not necessary to connect large masses of metal, such as engines, tanks, &c., to the walls, if they are entirely within the building. If, however, any conductor, such as a telegraph-wire, or a metallic supply-pipe for water or gas, comes into the building from without, the potential of this conductor may be different from that of the building, unless it is connected with the conducting shell of the building. Hence the water or gas supply-pipes, if any enter the building, must be connected to the system of lightning conductors; and since to connect a telegraph-wire with the conductor would render the telegraph useless, no telegraph from without should be allowed to enter a powder-mill, though there may be electric bells and other telegraphic apparatus within the building. I have supposed the powder-mill to be entirely sheathed in thick sheet copper. This, however, is by no means necessary in order to prevent any sensible electrical effect taking place within it, supposing it struck by lightning. It is quite sufficient to enclose the building with a net-work of a good conducting substance. For instance, if a copper wire, say No. 4, B. W. G. (0·238 inch diameter) were carried round the foundation of the house, up each of the corners and gables, and along the ridges, this would probably be a sufficient protection for an ordinary building against any thunderstorm in this climate. The copper wire may be built into the wall to prevent theft, but should be connected to any outside metal, such as lead

Fig. 20. ARRANGEMENT OF PROFESSOR CLERK MAXWELL'S LIGHTNING CONDUCTORS.

or zinc on the roof, and to metal rain-water pipes. In the case of a powder-mill, it might be advisable to make the network closer by carrying one or two additional wires over the roof and down the walls to the wire of the foundation. If there are water or gas-pipes which enter the building from without, these must be connected with the system of conducting wires; but if there are no such metallic connections with distant points, it is not necessary to take any pains to facilitate the escape of the electricity into the earth; still less is it advisable to erect a tall conductor with a sharp point in order to relieve the thunder-clouds of their charge.

'It is hardly necessary to add, that it is not advisable,

during a thunderstorm, to stand on the roof of a house so protected, or to stand on the ground outside, and lean against the wall.'

Prof. Clerk Maxwell, in a letter to Mr. Charles Tomlinson, F.R.S., the author of the 'Thunderstorm,' says : 'My plan is to convert a building into a closed conducting vessel by a sufficient number of wires enclosing it. For an ordinary house, a skeleton of its edge is quite enough. A *a* may be a zinc ridge, B *b* and C *c* water-gutters of zinc or iron ; but the pieces A B D, A C E, *a b d*, *a c e*, and the circuit D E *e d* should be of stout copper wire or rope, built into the wall as a security against theft, but connected to every other piece of metal on the outer surface of the house, and to every gas

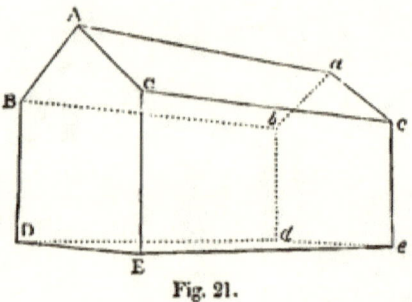

Fig. 21.

or water-pipe which enters the house from without, but *not* to any masses of metal wholly within the whole, unless this is desirable for other purposes.'

CHAPTER XIV.

ACCIDENTS AND FATALITIES FROM LIGHTNING.

THE accidents that occur annually from the effects of lightning are far greater in number and extent than is generally supposed. Although the art of protecting buildings by means of lightning conductors was discovered some hundred and twenty-seven years ago, and it is now one hundred and eleven years since, in 1768, Benjamin Franklin's 'lightning rods' were first set up over the dome of Saint Paul's Cathedral, yet the application of this great discovery is by no means general. At least one-half, and perhaps two-thirds, of all the public buildings, including the churches and chapels, of Great Britain and Ireland, are without any protection against lightning. As to private houses, it may safely be affirmed that not five out of every thousand are fitted with lightning conductors. It is well known that the amount of property annually destroyed by lightning in this country is very great, though it is, very naturally, impossible to form any accurate, or even approximate estimate of it. With regard, however, to the number of deaths from the same cause, certain statistics do exist, although many of them are notoriously imperfect. According to the 'Fortieth Report of the Registrar-General,' issued in July 1878, and former reports, the number of deaths from lightning in England and Wales was as follows in each of the nine years from 1869 to 1877 :—

Years	Males	Females	Total
1869	5	2	7
1870	13	6	19
1871	23	5	28
1872	35	11	46
1873	17	4	21
1874	25	—	25
1875	14	3	17
1876	15	4	19
1877	—	—	12
Total	147	35	194

The official returns of the number of deaths from lightning, as given by the English Registrar-General, are admittedly incomplete. In Prussia, where the registration of the causes of death is most rigidly enforced by law, and, in consequence, is far more accurate than in England, there were one thousand and four persons reported as killed by lightning in the nine years from 1869 to 1877. According to the official report issued by Dr. Ernst Engel, Director of the Statistical Bureau of Berlin, the number of lives lost each year was as follows:—

Years	Males	Females	Total
1869	47	32	79
1870	59	43	102
1871	56	47	103
1872	50	35	85
1873	61	50	111
1874	58	49	107
1875	92	48	140
1876	59	47	106
1877	105	66	171
Total	587	417	1,004

The population of Prussia is somewhat larger than that of England and Wales—25¾ millions against 24½ millions—but on the other hand, thunderstorms are less frequent there than with us. Altogether it will be rather under than over the mark to say that as many persons are killed by lightning in England as in Prussia, the loss amounting, on the average, to over one hundred every year.

Of the deaths by lightning in France, Mons. Boudin some

years ago collected statistics which showed that during the thirty years beginning in 1834 and ending in 1863, two thousand and thirty-eight people were struck dead by lightning in that country. During the last ten years of this period, the deaths amounted to eight hundred and eighty, and of these only two hundred and forty-three were females. In connection with this it is a noticeable fact that when a lightning stroke falls upon a crowd, it almost invariably causes more fatalities among the men than the women.

In the following tables are given statistics of deaths and accidents from lightning in the various countries referred to.

In the case of the United States, the Chief of the Bureau of Statistics writes that no record of deaths or fires caused by lightning is kept—a somewhat curious admission on the part of such a practical and methodical country. A similar reply has been received from the authorities in Spain.

CAS DE MORT, OCCASIONNÉS PAR LA FOUDRE, DANS LES 49 GOUVERNEMENTS DE LA RUSSIE EUROPÉENNE, SANS COMPTER LA FINLANDE ET LES GOUVERNEMENTS DU CI-DEVANT ROYAUME DE POLOGNE.

ANNÉES	HOMMES	FEMMES
1870	261	139
1871	260	167
1872	404	216
1873	300	179
1874	227	117
Total (en cinq années)	1,452	818
De ce nombre dans les villes	75	34
„ „ „ villages	1,377	784

INCENDIES, OCCASIONNÉS PAR LA FOUDRE, DANS LES 49 GOUVERNEMENTS DE LA RUSSIE EUROPÉENNE, SANS COMPTER LA FINLANDE ET LES GOUVERNEMENTS DU CI-DEVANT ROYAUME DE POLOGNE.

ANNÉES	DANS LES VILLES	DANS LES VILLAGES
1870	11	571
1871	23	767
1872	28	1,217
1873	19	908
1874	12	636
Total	93	4,099

Dans le gouvernement de Cherson les villes Odessa et Nicolaev ne sont pas comprises, à cause du manque de renseignements.

The returns from Russia, which include the years 1870, 1871, 1872, 1873, and 1874, are here printed as they were received from the President of the Commission for Statistics at St. Petersburg.

The returns from Sweden, extending as they do over a period of more than sixty years, are highly interesting. In this case the difference in the number of men and women killed is not so noticeable as in other countries:—

DEATHS BY LIGHTNING IN SWEDEN.

Year	Total	Of which					
		Under 10 years old	Over 10 years old	Males	Females	In the country	In the towns
1877	8	—	8	4	4	7	1
1876	14	2	12	6	8	14	—
1875	16	—	16	10	6	16	—
1874	9	—	9	6	3	9	—
1873	14	1	13	7	7	13	1
1872	26	2	24	10	16	25	1
1871	6	—	6	2	4	5	1
1870	9	1	8	5	4	9	—
1869	7	1	6	3	4	7	—
1868	14	—	14	11	3	14	—
1867	5	—	5	3	2	5	—
1866	26	2	24	8	18	25	1
1865	13	2	11	7	6	13	—
1864	5	1	4	2	3	4	1
1863	4	—	4	3	1	4	—
1862	12	2	10	10	2	11	1
1861	15	—	—	—	—	15	—
1860	7	—	—	4	3	7	—
1859	22	4	18	12	10	20	2
1858	18	—	18	11	7	17	1
1857	6	1	5	3	3	6	—
1856	6	1	5	3	3	6	—
1855	25	2	23	16	9	25	—
1854	5	—	5	4	1	5	—
1853	8	1	7	4	4	8	—
1852	15	4	11	7	8	14	1
1851	9	—	9	7	2	9	—
1850	9	3	6	6	3	9	—
1849	11	1	10	4	7	10	1
1848	5	—	5	1	4	5	—
1847	10	—	10	3	7	10	—
1846	21	1	20	14	7	21	—
1845	16	4	12	10	6	14	2
1844	11	—	11	9	2	10	1
1843	2	—	2	—	2	2	—
1842	7	1	6	5	2	7	—
1841	7	1	6	5	2	7	—
1840	2	—	2	1	1	2	—
1839	22	4	18	17	5	22	—
1838	11	1	10	9	2	11	—
1837	5	2	3	3	2	5	—

ACCIDENTS AND FATALITIES FROM LIGHTNING. 173

DEATHS BY LIGHTNING IN SWEDEN (continued).

Year	Total	Of which					
		Under 10 years old	Over 10 years old	Males	Females	In the country	In the towns
1836	4	—	4	2	2	4	—
1835	5	1	4	3	2	5	—
1834	36	4	32	21	15	36	—
1833	7	1	6	6	1	7	—
1832	5	—	5	1	4	5	—
1831	7	1	6	3	4	7	—
1830	5	—	—	5	—	—	—
1829	10	—	—	6	4	—	—
1828	9	—	—	6	3	—	—
1827	5	—	—	4	1	—	—
1826	11	—	—	6	5	—	—
1825	6	—	—	3	3	—	—
1824	6	—	—	4	2	—	—
1823	5	—	—	3	2	—	—
1822	8	—	—	4	4	—	—
1821	4	—	—	3	1	—	—
1820	15	—	—	8	7	—	—
1819	32	—	—	17	15	—	—
1818	10	—	—	4	6	—	—
1817	4	—	—	2	2	—	—
1816	7	—	—	3	4	—	—

DEATHS BY LIGHTNING IN BADEN.

Year	Males	Females	Total
1874	3	—	3
1875	3	5	8
1876	2	7	9
Total	8	12	20

FIRES THROUGH LIGHTNING IN BAVARIA.

Right side of Rhine.

Year	Total	Year	Total	Year	Total
1843–44	24	1853–54	38	1863–64	59
1844–45	30	1854–55	47	1864–65	90
1845–46	54	1855–56	70	1865–66	48
1846–47	25	1856–57	66	1866–67	100
1847–48	27	1857–58	56	1867–68	140
1848–49	26	1858–59	60	1868–69	86
1849–50	30	1859–60	50	1869–70	79
1850–51	32	1860–61	64	1870–71	115
1851–52	44	1861–62	63	1871–72	107
1852–53	60	1862–63	80	1872–73	170

Left side of Rhine.

Year	Total
1870	6
1873	36

174　　　　　　LIGHTNING CONDUCTORS.

Austria. List of Damages by Fire through Lightning.

Year	Total Fires	Austria, Western Europe	Austria, Eastern Europe	Salzburg	Styria	Kärnten	Illyria	Coastland	Tyrol	Bohemia	Mähren	Silesia	Galicia	Bockovina	Dalmatia	Total
								Of which those through lightning are:—								
1870	4,171	20	16	1	14	3	4	2	2	58	15	—	26	—	—	161
1871	4,283	9	26	1	9	5	2	3	1	53	8	1	34	4	—	156
1872	5,265	11	26	4	22	5	12	3	2	45	14	7	56	5	1	223
1873	5,500	11	16	3	30	4	12	1	11	68	18	7	42	3	3	249
1874	5,244	15	24	—	22	5	9	—	8	79	15	5	53	5	—	250
1875	4,529	17	34	—	19	4	10	2	7	68	19	8	62	—	—	250
1876	5,001	18	13	1	22	5	5	1	1	59	15	—	48	—	—	188
1877	6,135	21	23	3	23	8	8	—	7	63	19	6	43	1	—	225
Total	40,128	122	178	13	181	39	62	12	39	513	123	34	364	18	4	1,702

DEATHS BY LIGHTNING IN WURTEMBERG.

Year	Total
1841–42	26
1851–60	117

DEATHS BY LIGHTNING IN SWITZERLAND.

	Males	Females	Total
1876	2	1	3
1877	21	9	30
Total	23	10	33

The data given here is necessarily incomplete, although much trouble has been taken in obtaining it. Many countries keep no separate records of deaths and accidents from lightning, and those kept by others are often meagre and untrustworthy. Still the statistics given are sufficient to prove that lightning constitutes no unimportant factor among the dangers that threaten the safety of human life. The apathy with which the danger is regarded by most people is simply astounding: very few make any effort to protect themselves or their houses against it, although during certain months of the year it is almost impossible to take up a newspaper that does not contain an account of some fatality or casualty from the effects of a thunderstorm. The long roll of accidents appended to this chapter shows only too clearly the enormous amount of damage that has arisen—and is continually arising—from this source. Public buildings fare little better than private houses. Even some of the first cathedrals of England have no lightning conductors whatever, while others, supplied with them, are insufficiently protected, as is apparent to any competent observer. A glaring instance of the absence of protection against lightning is to be found at Windsor Castle. It is a fact that several portions of this splendid palace, among them St. George's chapel, and the adjoining Belfry Tower, are entirely unprovided with lightning conductors. On other parts of the castle a few conductors are placed, but clearly not enough. It is needless to say that, speaking only of St. George's chapel and the

Belfry Tower, these beautiful buildings, constantly touched by the storm-clouds that sweep up the valley of the Thames, are liable at any moment to destruction or great damage.

Thomas Fuller, in his 'Church History of Britain,' states that—

'There was scarce a great abbey in England which (once, at the least) was not burnt down with lightning from Heaven. 1. The Monastery of Canterbury, burnt anno 1145, and afterwards again burnt anno 1174. 2. The abbey of Croyland, twice burnt. 3. The Abbey of Peterboro, twice set on fire. 4. The Abbey of Mary's in York, burnt. 5. The Abbey of Norwich, burnt. 6. The Abbey of St. Edmondsbury, burnt and destroyed. 7. The Abbey of Worcester, burnt. 8. The Abbey of Gloucester. 9. The Abbey of Chichester, burnt. 10. The Abbey of Glastonbury, burnt. 11. The Abbey of St. Mary in Southwark, burnt. 12. The Church of the Abbey of Beverley, burnt. 13. The steeple of the Abbey of Evesham, burnt.'

Even in those cases where in modern times lightning conductors have been applied to buildings, they are very often improperly fixed in the first instance, or, having once been put up, are never examined or tested with the view of ascertaining their constant efficiency. Several accidents owing their origin to one or the other of these causes have occurred quite recently. In May 1879, the church at Laughton-en-le Morthen was struck by lightning and damaged in the manner described in Chapter XIII. The spire was fitted with a corrugated copper tube conductor the joints of which were made by screws and coupling-pieces, but there was no metallic contact between the lengths; the conductor was insulated from the building; and the earth-contact was obtained by bending the end of the tubing, and inserting it about twelve inches deep in dry loose rubbish! Such a conductor is worse than useless. If it had been examined by a competent person, it must at once have been utterly condemned. In June 1879, a disastrous result followed the use of a similar conductor erected upon a private house near

Sheffield. In this case the corrugated tube forming the conductor contained too little metal, and it was insulated from the building. The examples show the necessity of leaving the design and erection of lightning conductors to those persons who have made a thorough study of the subject, since the work is by no means so free from complexity as is commonly supposed.

Fig. 1. ST. GEORGE'S CHURCH, LEICESTER.

Figs. 1 and 2 show the tower and spire of St. George's Church, Leicester, after being severely damaged by lightning on August 1st, 1846. The storm, during the course of which it was struck, was very violent, of prolonged duration, and accompanied by torrents of rain and hail. Mr. Charles Tomlinson, F.R.S., in his work on 'The Thunderstorm,' thus describes the catastrophe:—

'It was at five minutes past eight, after one or two peals

of unusual distinctness, that the church of St. George was
struck with a report resembling the discharge of cannon, and
with a concussion of the air which shook the neighbouring houses, and extinguished a lamp burning at the entrance
of the News-room, many hundred feet distant. The Sexton
had gone into the church, as usual, to toll the eight o'clock
bell; but was so terrified by the "fire-balls" that he saw in
the sky, and by the fact that once or twice the clapper
struck the side of the bell without his agency, that he made
his work as short as possible, and had just gone out and
locked the churchyard gate when the stroke fell. Two of
the spectators of this awful event were Captain Jackson and
the Rev. R. Burnaby, the rector of the parish, who both
described the flash as a vivid stream of light, followed by a
red and globular mass of fire, and darting obliquely from
the north-west, with immense velocity, against the upper
part of the spire. For the distance of forty feet on the
eastern side, and nearly seventy on the west, the massive
stone-work of the spire was instantly rent asunder and laid
in ruins. Large blocks of stone were hurled in all directions,
broken into small fragments, and in some cases, as there is
every reason to believe, reduced to powder. One fragment
of considerable size was hurled against the window of a
house three hundred feet distant, shattering to pieces the
woodwork, as well as fourteen out of the sixteen panes of
glass, and strewing the room within with fine dust and
fragments of glass. It has been computed that a hundred
tons of stone were on this occasion blown to a distance of
thirty feet in three seconds. In addition to the shivering of
the spire, the pinnacles at the angles of the tower were all
more or less damaged, the flying buttresses cracked through
and violently shaken, many of the open battlements at the
base of the spire knocked away, the roof of the church completely riddled, the roofs of the side-entrances destroyed,
and the stone staircases of the gallery shattered. The top
of the spire, when left without support beneath, fell per-
pendicularly downwards, inside the steeple, causing much

ACCIDENTS AND FATALITIES FROM LIGHTNING.

devastation in its descent. Falling through the uppermost storey, and carrying along with it the bell and its solid supports, the ruined spire entered the room containing the clock, dashed the works to pieces, and penetrating the strong and well-supported floor, descended with additional momentum through the third and fourth floors (the latter being that just deserted by the prudent sexton), and reached the paved vestibule with so furious a shock as to drive in a portion of the strong foundation-arch, by which the weight of the whole tower was supported. On looking upward from the scene of ruin in this vestibule, the tower appeared like a well, so small were the vestiges of its various storeys.'

After minute examination, it was evident that the course of the lightning had been nearly as follows:—'The flash first struck the gilded vane, marks of lightning being perceptible between its bevelled edges. After traversing the vane and spindle, and the terminating iron supports, the only path left for the fluid was through a series of iron cramps, separated by masses of sandstone; and here it was that the explosion commenced—the stone being torn and hurled aside as it came in the path of the lightning to the lowest lead lights of the spire. Most of these iron cramps were found to be powerfully magnetic; and one of them, eight weeks afterwards, sustained a very considerable brush of steel filings at its edges. The lattices of the lights on three sides of the spire were little injured; but on the fourth side the stone-

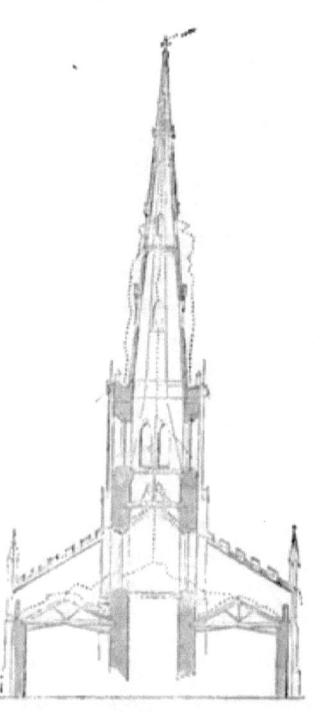

Fig. 2. ST. GEORGE'S CHURCH, LEICESTER.

work was shattered, and the lattice singularly twisted and partially fused. Here, it appears, another violent explosion took place, and the lightning diverging struck the north-west pinnacle, attracted apparently by the copper bolt by which the stones were held together. It also struck the large cast-iron pipe on the other side of the spire, reaching from the tower-battlements to the roof of the church; and during its passage down the pipe, and at an inequality in the surface of the metal, it displayed the most extraordinary expansive force, bursting open and scattering to a distance portions of metal of great solidity and weight. From the leadwork of the roof the lightning was conducted to the leaden gutters, and so finally to the earth.

'The course of the remaining current in the interior of the tower was first to be traced on the lattices of the belfry, then in the clock-room, where the works of the clock were strongly magnetised, thence in at least three different directions to the outside of the tower. The external faces of the clock were not much altered, the hands were, however, slightly discoloured, and the blackened surfaces of the dials covered with streaks, as if smeared with a painter's brush. On quitting the dial faces on the northern and southern sides of the tower, the lightning evidently fell upon the leads of the side lobbies, and was finally carried off by the two iron pipes connecting their roofs with the earth. Both these pipes were chipped and injured, and one of them was perforated, as if by a musket-shot, a few inches from the ground. The edges of this fracture were found to possess magnetic power. Thus, besides the division of the current at the upper part of the spire, there was a second division in at least three directions from the clock-room and dial faces. The roof of the church throughout its whole extent showed signs of an extraordinary diffusion of the electric current; and in almost every place where one piece of metal overlapped another, a powerful explosion had evidently taken place.'

As far as is known the church was unprovided with any lightning conductor. The same storm produced most

ACCIDENTS AND FATALITIES FROM LIGHTNING. 181

disastrous effects in other parts of the Kingdom. Seven thousand panes of glass were broken by the hail in the Houses of Parliament; three hundred at the Police Office, Scotland Yard; other buildings in the metropolis suffered to

Fig. 3. WEST-END CHURCH, SOUTHAMPTON.

a similar extent, the glass in the picture gallery at Buckingham Palace being totally destroyed and the apartment flooded with water.

Fig. 3 shows the spire of the church at West End,

Southampton, which was struck by lightning on June 10, 1875. A large quantity of the stone-work was broken by

Fig. 4.   MERTON COLLEGE CHAPEL, OXFORD.

the passage of the electric discharge, and some of the pieces were thrown to a great distance.

On September 27, 1875, the tower of the chapel of Merton College, Oxford, was struck by lightning. The damage done was confined to the mutilation of one of the corner pinnacles and the displacement of fragments of some of the stonework which were thrown on to the leads and the pathway beneath. Some workmen were on the leads at the time, but fortunately were not hurt. The tower had lately been restored, and the scaffolding had only been removed a few days previous to the accident. A gentleman who had taken shelter from the storm in one of the workmen's sheds beneath the tower was startled by seeing fragments of stone falling from above; looking up, he discovered that the tower had been struck, and immediately informed the college authorities. On ascending the tower it was found that one of the eight crocketted pinnacles had been struck. This pinnacle occupied the south-western corner, and had been completely and cleanly severed from summit to base. Fortunately, the stonework displaced—which weighed about three hundredweight—was thrown on to the roof of the tower, a distance of twenty-five feet. The vane, slightly fused by the electric discharge, was found embedded in an upright position in the leads. The mouldings on the edges of the pinnacle were divided to the extent of four feet, and many of the stones were turned entirely round.

Fig. 5.

The tower, which was erected in 1424, and is one of the landmarks of Oxford, had not up to the time of the accident been provided with lightning conductors, but they have since been affixed to the building.

Fig. 5 shows the steeple of St. Bride's Church, Fleet Street, London, which was severely damaged by lightning on June 18, 1764. The spire of this steeple is built in four storeys, surmounted by an obelisk. These four storeys are braced together by means of horizontal iron bars; another iron bar, about twenty feet long and two inches square, passes through the upper part of the obelisk and supports the weather-cock and other ornamental work on the top; there is also a great deal of iron-work used generally in the construction of this part of the building, thus forming a complete series of discontinuous metallic masses. When the lightning struck the building it was received by the long iron bar which supported the vane; at the lower end of the bar the electric discharge, meeting with no metallic conductor, burst with great violence, shattering the stone on which the bar rested; the lightning then pursued its course to earth, leaping from one piece of metal to another and breaking the stone-work in its way. The last trace of it was found at the west window of the belfry, from whence it seems to have found a road to the earth. The damage sustained by the structure was so great that eighty-five feet of the spire had to be entirely rebuilt.

The edifice was afterwards attentively examined (as explained in a previous chapter) by Dr. Watson, a well-known electrician in those days, who reported to the Royal Society that the accident 'completely indicated the great danger of insulated masses of metal to buildings from lightning; and, on the contrary, evinced the utility and importance of masses of metal continued and properly conducted, in defending them from its direful effects. The iron and lead employed in this steeple, in order to strengthen and preserve it, did almost occasion its destruction; though, after it was struck by the lightning, had it not been for these materials keeping the remaining parts together, a great part of the steeple must have fallen.

Fig. 6 shows the condition of St. Michael's Church, at Black Rock, near Cork, after being struck by lightning on January 29, 1836. The damage done was entirely on the windward side of the steeple, caused, as is suggested in Mr. Tomlinson's work, by this side receiving the greatest quantity of rain, and so being rendered the 'line of least resistance,' but not a sufficiently good conductor to carry off the discharge to earth.

On Trinity Sunday, June 8th, 1879, a violent thunderstorm broke over the town of Wrexham about half-past three in the afternoon, during which the spire of St. Mark's Church was struck by lightning. A Sunday-school class was being held in the room at the base of the spire, and the teacher and five of his scholars were burnt, three of them severely, and one had his leg broken. Some of the stone-work of the spire was also displaced and thrown down. The spire was fitted with a copper conductor, but it was

Fig. 6.

of too small a calibre, and it is very doubtful whether the earth connection was all it should have been.

Many other cases might be enlarged upon, but enough have been given to prove the imperative need for a more general use of lightning conductors on all public and private buildings. Another equally important necessity is that lightning conductors should be erected on sound principles, and also be periodically examined and tested by some competent person.

PUBLIC BUILDINGS STRUCK BY LIGHTNING.

Date	Building	Damage	Authority
1589. July 16	Nicholas Tower, Hamburg	The tower burnt down	From original notices in Reimarus, Bl. 315
1670. June 29	Nicholas Church, Stralsund	Damaged.	Phil. Trans. v. 2084
1673. June 29	Pharr Church, Epperies, Hungary	Damaged.	Breslauer Samml. 1717, p. 64
1693.	Oundle Church	Set on fire	Phil. Trans. xvii. 710
1700. Oct. 9.	Principal church, Troies	Set on fire and shattered	Mém. de l'Acad. de Sc. Paris, 1700, p. 65
1708. July	All Hallows' Church, Colchester	Damaged.	Phil. Trans. 432
1711. May 20	Principal town tower in Bern; houses adjoining	Damaged.	Scheuchzer, Meteorol. Helv. p. 35
1711. May 23	The belfry of the church at Solingen	Set on fire	Scheuchzer, Meteorol. Helv. p. 28
1714. June 21	Elizabeth Tower, Breslau	Damaged.	Breslauer Samml. 1717, p. 68
1717. July 2.	Church at Seidenberg, near Zittau	Seven persons killed	Breslauer Samml. 1718, p. 1534
1718. April 14, 15	Twenty-four churches between Landerneau and St. P. de Léon, Brittany	Set on fire and shattered	Hist. de l'Acad. de Sc. Paris, 1719, p. 21
1718. Dec. 14	Church tower at Eutin	Set on fire	Breslauer Samml. 1718, p. 1968
1725. Dec. 18	Church tower, Winterthur	Lightning followed an accidental conductor, and resulted in melting it	Breslauer Samml. 1725, p. 166
1728. Aug. 25	Church tower, Mellingen, in Baden	Shattered.	Reimarus, Bl. 145
1731. July	Three villages near Geneva	Destroyed	Gent.'s Mag. p. 309
1732. Oct.	The Escurial at Madrid	Set on fire	Gent.'s Mag. p. 1034
1743. Aug.	Liberton Church, Scotland	Steeple destroyed	Gent.'s Mag. xiv. 450
1745. July 21	Tower of monastery, Bologna	Shattered. Lightning followed an accidental conductor, and melted it	Reimarus, Bl. 93
1746. May 21	Tower of the School Church, Halle	The ball on the tower bent, and other mechanical effects	Reimarus, 198
1747. Aug. 20	Tower of the College Church, Pluviers	Physio. and mechanical effects	Mém. de l'Acad. de Sc. Paris, 1748, p. 572
1748. May 31	Top of a church tower, Witzendorf	Shattered and tore off the roof; melted and shattered accidental conductor	Hamburg Magazine, ix. 301

ACCIDENTS AND FATALITIES FROM LIGHTNING. 187

Public Buildings struck by Lightning (*continued*).

Date	Building	Damage	Authority
1750. Feb. 5.	Church tower, Danbury, Essex	Set on fire	Phil. Trans. xlvi. 611
1750. Spring.	Tower of Dutch Church, New York	Lightning followed an accidental conductor, which was shattered, and caused other mechanical effects	Franklin, Experiments and Observations, xv. 180
1751. June 6	Tower of church, South Moulton, Devonshire	Lightning followed an accidental conductor, which was shattered, and caused other mechanical effects	Phil. Trans. xlvii. 330
1752. June 19	Church tower, Alfwa, Sweden	Tower damaged; several persons injured	Schwed. Abh. xv. 80
1753. Mar. .	Darlington Church .	Much damaged	Gent.'s Mag. xxiii. 145
1753. Oct. .	Church of Les Filles de St. Sacrament, Naples	Reduced to ashes	Gent.'s Mag. xxiii. 487
1754. June 16	Belfry of Newbury Church	Point of spire shattered, accidental conductor melted, and other damage	Phil. Trans. xlix. 307
1755 . . .	Danish Church, Wellclose Square	Clock damaged	Phil. Trans. xlix. 208
1755. Dec. .	St. Aubin Church, Lorraine	Much damaged	Gent.'s Mag. xxv. 42
1757. Jan. .	Lostwithiel Church, Cornwall	Much damaged	Gent.'s Mag. xxviii. 427
1757. Nov. .	ChristChurch, Dublin	Much damaged	Gent.'s Mag. xxvii. 527
1759. April .	Great Billing Church, Northamptonshire	Steeple destroyed	Ann. Reg. ii. 84
1759. May .	Portsmouth Church, New Hampshire	Much damaged	Gent.'s Mag. xxix. 355
1759. June 10	Jacob Church, Aumale	Several persons injured	Reimarus, Bl. 158
1760. July 16	Church, Altona .	Lightning struck the copper covering on the top of spire, followed accidental conductors, and melted them	Reimarus, 59
1761. June .	Shifnal Church, Norfolk	Greatly damaged	Ann. Reg. iv. 136
1761. July .	Ludgvan Church, near Penzance	Greatly damaged	Ann. Reg. iv. 142
1763. Mar. .	Harrow Church	Set on fire	Gent.'s Mag. xxiii. 142
1763. Mar. .	Salisbury Cathedral	Damaged	Gent.'s Mag. xxiii. 143
1763. Mar. .	Southam Church, Warwickshire	Damaged	Gent.'s Mag. xxxiii. 142
1764. June 18	St. Bride's Church, London	Spire struck and much damaged	Phil. Trans. liv. 227
1765. Aug. .	Bicester Church .	Much damaged	Gent.'s Mag. xxxv. 391

188   LIGHTNING CONDUCTORS.

PUBLIC BUILDINGS STRUCK BY LIGHTNING (*continued*).

Date	Building	Damage	Authority
1766. July	Skipton-in-Craven Church	Much damaged	Ann. Reg. ix. 118
1766. Aug.	St. Mary's Church, Bury St. Edmunds	Much damaged	Ann. Reg. ix. 122
1767. April	Provence, France	Three churches set on fire	Ann. Reg. x. 81
1767. May	Mentz Cathedral	Set on fire	Ann. Reg. x. 92
1767. Aug. 6	Nicholas Tower, Hamburg	Lightning followed accidental conductors, and partly melted them	Reimarus, Bl. 201
1767. Sept.	Genoa	Several churches damaged	Ann. Reg. x. 126
1768. Aug. 21	Church tower in Alem	Damaged. Several persons injured	Haarlem Verh. xiv. 34
1770. Feb. 18	St. Keverns Church, Cornwall	Damaged. Several persons injured	Hemmer, Act. Acd. Palat. iv. 37
1771. Feb. 2	Nicholas Church, Kiel	Lightning followed accidental conductor, and left traces	Ackermann's notice, Kiel, 1772
1772. Mar.	St. Paul's Cathedral, London	Lightning followed accidental conductor, and left traces	Arago, iv. 88
1773. April	Lighthouse at Villafranca, Nice	Destroyed	Gent.'s Mag. xliii. 246
1773. June	Rhichenback, Saxony	Town reduced to ashes	Ann. Reg. xvi. 115
1773	St. Peter's Church, London	Shattered the tower roof	Phil. Trans. lxv. 336
1774. Aug.	Buckland Church, near Dover	Damaged	Ann. Reg. xvii. 140
1775. Feb.	St. Colomb Church, Cornwall	Much damaged	Ann. Reg. xviii. 91
1775. June 27	A church in Munich	Tower injured	Epp. 90
1776. Aug.	Cuckfield Church, Suffolk	Much damaged	Ann. Reg. xix. 170
1778. April 15	Church in Altona	Metal melted	Reimarus, Bl. 64
1780. Sept.	Church of the Holy Spirit, Hamburg	Injured	Reimarus, N.B. 47
1780. Oct.	Hammersmith Church	Much damaged	Ann. Reg. xxiii. 230
1783. July	Ashbourne Church, Derbyshire	Steeple demolished	Gent.'s Mag. liii. 707
1783	St. Mary's, Leicester	Steeple demolished	
1786. June 26	Church in Wachenheim	Shattered. People injured	Act. Acad. Theod. Palat. vi. 332
1787. June	St. Mary's Church, Grenoble	Much damaged	Gent.'s Mag. lvii. 820
1787. June	Vendamir Church, Vercovia	Several persons killed	Gent.'s Mag. lvii. 820
1787. June	St. Gregorius Church, Prague	Set on fire	Gent.'s Mag. lvii. 820
1787. June	Cranbrook Church	Much damaged	Gent.'s Mag. lvii. 824
1789. May	Pforzheim Church	Entirely consumed, with thirty adjoining buildings	Gent.'s Mag. lix. 754
1789. June	Barnewell Church, near Oundle	Damaged	Gent.'s Mag. lix. 665
1790. Dec.	Beckenham Church	Set on fire	Ann. Reg. xxxii. 229

ACCIDENTS AND FATALITIES FROM LIGHTNING.

PUBLIC BUILDINGS STRUCK BY LIGHTNING (*continued*).

DATE	BUILDING	DAMAGE	AUTHORITY
1790. Dec.	Horsham Church	Set on fire	Ann. Reg. xxxii. 229
1791. Jan.	Ashton-under-Lyne Church	Much damaged	Ann. Reg. xxxiii. 3
1791. Oct.	Rainham Church	Much damaged	Gent.'s Mag. lxi. 1050
1795. June	Castor Church	Much damaged	Gent.'s Mag. lxv. 517
1795. Dec. 25	Church in Bergen, Norway	Set on fire	Gilb. Ann. xxix. 176
1797. July	Grantham Church	Damaged	Gent.'s Mag. lxviii. 104
1797. Aug.	Caldecot Church, Rutland	Spire much damaged	Gent.'s Mag. lxvii. 817
1801. July	Corby Church	Damaged	Gent.'s Mag. lxxi. 659
1804. Mar.	St. Gertrude Church at Nevelles	Burnt by lightning	Gent.'s Mag. lxxiv. 368
1804. Mar.	St. Maria at Oudenard in Flanders	Burnt by lightning	
1804. June	Edenham Church, Lincoln	Damaged	Ann. Reg. xlvi. 394
1804. June	Hanslope Church, Bucks	Spire destroyed	Ann. Reg. xlvi. 395
1806. July	Sunbury Church, Middlesex	Damaged	Ann. Reg. xlviii. 426
1807	Montvilliers Church, France	Damaged	Howard's Climate of London, ii. 20
1810. July	Attercliffe Chapel	Much damaged	Gent.'s Mag.
1811. June	Ashford Church	Much damaged	Gent.'s Mag. lxxxi. 584
1811. Dec.	Ledbury Parish Church	Damaged	Gent.'s Mag. lxxxi. 650
1812	St. Pelverin Church, Department of the Loire	Set on fire and burnt to the ground	Howard's Climate of London, ii. 165
1813	Bridgwater Church	Spire destroyed	Howard's Climate of London, ii. 222
1813	Weston Zoyland Church	Tower much damaged	Howard's Climate of London, ii. 222
1814. Nov.	Thackstead Church, Essex	Much damaged	Gent.'s Mag. lxxxiv. 491
1815	The steeples of many churches in Belgium, in places far distant from one another	Struck and set on fire nearly at the same hour	Howard's Climate of London, ii. 259
1816. July	Worschetz, county of Temeswar	Church and the town greatly damaged	Ann. Reg. lviii. 102
1816. Oct.	Moselle Church	Damaged	Ann. Reg. lviii. 161
1817. Mar.	St. Paulinas Church, Germany	Set on fire	Ann. Reg. lix. 15
1819. Jan.	St. Martin's Church, Guernsey	Much damaged	Ann. Reg. lxi. 5
1819. July	Sedgeford Church, Lynn	Much damaged	Ann. Reg. lxi. 50

Public Buildings struck by Lightning (*continued*).

Date	Building	Damage	Authority
1821. May 7 .	Tower of Katherine's Church, Gross-Selten	Church burned	Gilb. Ann. lxviii. 224
1821. May 7 .	Wooden Tower of Katherine Church, Tischendorf	Tower burned	Gilb. Ann. lxviii. 224
1821. May 8 .	Church at Carlsruhe	Damaged	Gilb. Ann. lxviii. 224
1821. Apl. .	Redcliffe Church, Bristol	Much damaged	Gent.'s Mag. xci. 367
1822. Jan. 15	Church at Gerstetten	Damaged	Wurtemberger Jahreshafte, xi. 463
1822. June .	North Luffenham Church, Rutland	Much damaged	Gent.'s Mag. xcii. 636
1822. Aug. .	Church at Chatham	Spire ripped open	Tomlinson's Thunderstorm, p. 165
1822. Sept. .	Rouen Cathedral	Set on fire	Tomlinson's Thunderstorm, p. 165
1822. Oct. .	St. Peter's Church, Venice	Reduced to ruins	Gent.'s Mag. xcii. 553
1823 . . .	Kemble Church, Wilts	Spire destroyed	Howard's Climate of London, iii. 135
1823. Feb.	Shaugh Church, near Plymouth	Tower struck and much shattered. An iron conductor had been erected about two years before, but this had rusted and gone to decay	Tomlinson's Thunderstorm, p. 165
1824. July 10	Church at Simmerfeld	Damaged	Würtemberger Jahreshafte, xi. 463
1824. Nov. .	Charles Church, Plymouth	Steeple struck, and the small brass rod erected as a lightning conductor knocked to pieces	Tomlinson's Thunderstorm, p. 165
1825—about .	Torrington Church, North Devon	Tower and steeple ruined. They had to be rebuilt	Tomlinson's Thunderstorm, p. 165
1826. June .	Alphington Church, near Exeter	Much damaged	Ann. Reg. 1826, p. 97
1827 . . .	Pailant Church, Chichester	Considerably damaged	Howard's Climate of London, iii. 259
1827. Jan. 11	Church Tower, Bussen	Set on fire, although covered with snow.	Würtemberger Jahreshafte, xi. 463
1828. Apl. .	Edlesborough Church	Set on fire	Gent.'s Mag. xcviii. 358
1828. June .	Kingsbridge Church, Devon	Steeple rent, and other damage	Tomlinson's Thunderstorm, p. 165
1828. Oct. .	Kilcoleman Church, co. Mayo	Spire destroyed	Ann. Reg. p. 131
1830. July .	Independent Chapel, Edgworth Moor, near Bolton	Damaged	Ann. Reg. p. 101

PUBLIC BUILDINGS STRUCK BY LIGHTNING (*continued*).

DATE	BUILDING	DAMAGE	AUTHORITY
1830. Aug.	Marlborough Church, near Kingsbridge, Devon	Tower and church severely damaged	Tomlinson's Thunderstorm, p. 166
1831. Feb.	Kilmichael Church, Glassire	Much damaged	Ann. Reg. p. 39
1833. Aug.	Strasburg Cathedral	Much damaged	Builder, ii. 39
1835. May 16	Church Tower, Endersbach	Much shattered	Würtemb. Jahreshafte, xi. 465
1835. June	Durham Cathedral	Western tower damaged	Ann. Reg. p. 94
1836. Jan.	Black Rock, near Cork	Spire demolished	Tomlinson's Thunderstorm, p. 166
1836. Nov.	Christ Church, Doncaster	The spire shattered and the church greatly injured. The roof was smashed in, and the churchyard presented a scene of ruin and devastation. The spire was surmounted by a ball of glass to keep off the lightning!	Tomlinson's Thunderstorm, p. 166
1837. June	Hoo Church, Kent	Set on fire	Gent.'s Mag. N.S. viii. p. 80
1839. Jan 8.	Church tower in Hasselt	Damaged	Arago, Notiz, 125
1841. Jan.	Spitalfields, London	Spire rent, and other damages	Tomlinson's Thunderstorm, p. 166
1841	Streatham	Spire nearly destroyed, and church set on fire	Tomlinson's Thunderstorm, p. 166
1841. May 10	Walton Church, Stafford	Spire destroyed	Tomlinson's Thunderstorm, p. 166
1841. Aug. 24	St. Michael's, Liverpool	Beautiful spire shattered, and clock injured	Tomlinson's Thunderstorm, p. 166
1841. Aug. 24	St. Martin's, Liverpool	Spire shattered, and other damage	Tomlinson's Thunderstorm, p. 166
1841	Wolverhampton Parish Church	Set on fire	Annals of Electricity, vi. 504
1841	Spitalfields Church	Steeple damaged	Annals of Electricity, vi. 504
1842. April 24	Brixton Church, London	Dome and building much rent	Tomlinson's Thunderstorm, p. 166
1842. July 28	St. Martin's, London	Spire shattered; cost of repair, 1,500*l*.	Tomlinson's Thunderstorm, p. 166
1843. April 25	Exton Church, Rutland	Spire destroyed; church set on fire and nearly destroyed	Tomlinson's Thunderstorm, p. 166
1843. May 25	St. Mark's, Hull	Slightly damaged	Tomlinson's Thunderstorm, p. 166
1843. Oct.	North Huish, near Modbury, Devon	Steeple shattered	Tomlinson's Thunderstorm, p. 166
1844. Mar.	Oving Church, near Chichester	Spire damaged	Tomlinson's Thunderstorm, p. 166

PUBLIC BUILDINGS STRUCK BY LIGHTNING (*continued*).

DATE	BUILDING	DAMAGE	AUTHORITY
1844 . . .	St. Clement's, London	Clock injured . .	Tomlinson's Thunderstorm, p. 166
1844. July .	Magdalen Tower, Oxford	One of the pinnacles damaged; staircase injured	Tomlinson's Thunderstorm, p. 106
1844. July 20	Stannington Church, near Sheffield	Seriously damaged	Tomlinson's Thunderstorm, p. 166
1846. June 14	Church near Chambrey	Damaged . .	Compt. Rend. xxiii. 153
1846. Aug. .	St. George's Church, Leicester	Spire destroyed .	Builder, iv. 395
1846. Aug. .	Dedham Church, Essex	Much damaged .	Builder, iv. 395
1846. Oct. .	Village of Schledorf, near Munich	Completely destroyed	Journal des Débats, Oct. 20, 1846
1847. June .	Her Majesty's palace, Osborne	One tower much damaged	Builder, vii. 291
1847. June .	Church in Thann	Much damaged .	Compt. Rend. xxix. 485
1847. Aug. .	Walton Church, Lincolnshire	Lightning entered at the belfry; one man killed, several injured	Tomlinson's Thunderstorm, p. 158
1849. July .	St. Saviour's, Southwark	Damaged . .	Ann. Reg. xci. 80
1850. May .	Norton - by - Gaulby Church	Spire much damaged	Builder, viii. 248
1850. May .	Little Stretton Church	Much damaged .	Builder, viii.
1850. Aug. .	Roman Catholic Church, York	Bell-turret shattered	Builder, viii. 405
1850. Oct. .	Keysoe Church .	Considerably damaged	Builder, viii. 509
1850. Nov. .	Cobridge Church, Potteries	Considerably damaged	Builder, viii. 533
1851. May .	St. Sepulchre's Church, Northampton	Much damaged	Builder, ix. 329
1851. May .	Edinburgh Assembly Hall	Much damaged	Builder, ix. 305
1851. June .	Boulogne Cathedral	Dome damaged .	Builder, ix. 415
1852. July 6 .	Ross Church, Hereford	Severely damaged .	Tomlinson's Thunderstorm, p. 166
1852. July .	Woolpit Church, Suffolk	Tower and spire destroyed	Builder, x. 492
1852. July .	Leighton Buzzard Church	Much damaged .	Builder, x. 492
1852 . . .	Exton Parish Church	Church nearly destroyed	Builder, xii. 575
1853. Jan. .	Derby Church . .	Much damaged .	Builder, xi. 28
1853. Jan. .	Parish Church, Eskdalemuir, Dumfries	Entirely destroyed .	Builder, xi. 43
1853. Feb. .	Lincoln Cathedral ,,	Struck north-west pinnacle of the broad tower; set on fire; narrowly escaped destruction	Tomlinson's Thunderstorm, p. 166
1853. July .	Skipton Church .	Much damaged	Builder, xi. 423
1853. July .	Hereford Old Parish Church	Slightly damaged	Builder, xi. 487

ACCIDENTS AND FATALITIES FROM LIGHTNING.

PUBLIC BUILDINGS STRUCK BY LIGHTNING (*continued*).

Date	Building	Damage	Authority
1853. Nov.	Chaddesley Corbett Church	Considerably damaged	Builder, xi. 704
1854. May	Hanwell Church	Spire much damaged	Builder, xii. 283
1854. May	Helpringham Church	Spire much damaged	Builder, xii. 200
1854. June	Ealing Church	Had a common conductor, which was fused; the church slightly damaged	Tomlinson's Thunderstorm, p. 167
1854. July	Ashbury Church	Had a common conductor, which was fused; church damaged, but not considerably	Tomlinson's Thunderstorm, p. 167
1854. July 19	Tower of Magdalen College, Oxford	Much damaged	Tomlinson's Thunderstorm, p. 167
1854. Aug.	National School Chapel, St. Mary, Ipswich	Three children killed, several injured	Ann. Reg. xcvi. 140
1855. May	Trinity Church, Southwark	Slightly damaged	Builder, xiii. 230
1855. May	St. Mark's, Myddelton Square	Considerably damaged	Builder, xiii. 230
1855. July	Holy Trinity Church, Brompton	Slightly damaged	Builder, xiii. 348
1855. July	St. Ebbe's Parish Church	Slightly damaged	Builder, xiii. 348
1856. Feb.	Chimney at Liverpool, 310 ft. high	Much damaged; struck at 20 yds. below the top	Tomlinson's Thunderstorm, p. 167
1856. June	Heminghrough Ch.	Much damaged	Builder, xiv. 348
1856. July	Clapton Church	Much damaged	Builder, xiv. 391
1856. July	Addlethorpe Church, Lincolnshire	Much damaged	Builder, xiv. 391
1856. July 14	Church of St. Ebbe, Oxford	Much damaged	Tomlinson's Thunderstorm, p. 167
1856. Aug.	Holy Trinity Church, Manchester	Much damaged	Builder, xiv. 461
1857. May	Parish Church, Wisborough, Sussex	Steeple set on fire	Tomlinson's Thunderstorm, p. 167
1857.	Walgrave Church	Damaged	Tomlinson's Thunderstorm, p. 167
1857. May	Wargrave Church, Twyford	Pinnacle destroyed	Tomlinson's Thunderstorm, p. 167
1857. Aug.	Tower of Windsor Castle	Four tons of parapet demolished	Tomlinson's Thunderstorm, p. 167
1857.	Independent Chapel, Portsmouth	Set on fire	Tomlinson's Thunderstorm, p. 167
1857. Aug.	St. Michael's Church, Stamford	Pinnacle demolished	Tomlinson's Thunderstorm, p. 167
1857.	Trinity Church, Southwark	Struck during service	Tomlinson's Thunderstorm, p. 167
1857. Aug.	A gasometer at the Chartered Gas Co.'s works, St. Luke's	Struck, and gas ignited	Builder, xv. 488
1858. July	The monument to Dugald Stuart at Edinburgh	Slightly injured	

LIGHTNING CONDUCTORS.

Public Buildings struck by Lightning (*continued*).

Date	Building	Damage	Authority
1858. July	Peak Hall, near Stoke-on-Trent	Church struck; roof damaged, walls seriously fractured, and organ injured	Tomlinson's Thunderstorm, p. 167
1862. May	Mashbury Church, Essex	Set on fire	Builder, xx. 391
1862. May	Bampton Parish Church	Much damaged	Builder, xx. 391
1862. May	Rainham Parish Church, Kent	Damaged	Builder, xx. 391
1862. July	Tackley (near Woodstock) Parish Church	Much damaged	Building News, 1862, p. 77
1863. Feb.	Dunoon Church, Scotland	Nearly destroyed	Builder, xxi. 140
1863. June	St. Paul's Church, Manchester	Considerably damaged	Building News, 1863, p. 457
1864. Sept.	St. Mary, York	Slightly damaged	Builder, xxii. 691
1865. Jan.	St. Lawrence, Nuremberg, Bavaria	Much damaged	Builder, xxiii. 53
1865. July	St. Mary's Church, Stamford	Much damaged	Builder, xxiii. 526
1865. July	St. Botolph Church, Boston	Much damaged	Builder, xxiii. 526
1865. July	Roman Catholic Chapel, Colchester	Much damaged	Builder, xxiii. 526
1867. Sept.	Sutton-in-Ashfield Church, Nottinghamshire	Spire destroyed	Builder, xxv. 605
1867. Sept.	St. Pé-Saint-Simon Church, France,	**Much** damaged	Builder, xxv. 684
1867. Sept.	Sanzet Church	Set on fire	Builder, xxv. 684
1868. May	St. Paul's Church, Little Chester, Derby	Much damaged	Builder, xxvi. 340
1868. June	St. Stephen's, Southwark	Slightly damaged	Builder, xxvi. **433**
1868. June	Temporary Congregational Church, Buckhurst Hill	Set on fire	Builder, xxvi. **433**
1868. June	Victoria Tower, Houses of Parliament	**Slightly damaged**	Builder, xxvi. 416
1868. June	Morville **Church**, Shropshire	Much damaged	Builder, xxvi. 416
1868. June	School, Furze **Hill**, Brighton	Much damaged	Builder, xxvi. 416
1868. June	Church, Shanghai	Destroyed	Builder, xxvi. 416
1870	St. Saviour's, Southwark	One pinnacle destroyed and church damaged	Builder, xxviii. 604
1870	**Rotherfield Church**	Considerably damaged	Builder, **xxviii. 604**
1871. June	Hethersett Church	Much damaged	Builder, xxix. 450
1871. June	St. John's Church, Bury St. Edmunds	Slightly damaged	Builder, xxix. 450
1871. July	St. Margaret's Church, King's Lynn	Much damaged	Ann. Reg. p. 72
1871. July	Cromer Church	Damaged	Ann. Reg. p. 72

ACCIDENTS AND FATALITIES FROM LIGHTNING.

PUBLIC BUILDINGS STRUCK BY LIGHTNING (continued).

Date	Building	Damage	Authority
1871. Sept.	Congregational Church, Terre Haute, Ind., U.S.	Considerably damaged	Scientific American, xxv. 161
1872. Jan.	St. Mary's Church, Crumpsall, Manchester	Set on fire and destroyed	Builder, xxx. 51
1872. June	Baptist Chapel, Wem	Slightly damaged	Builder, xxx. 511
1872. June	St. Mary's Church, Beeston, Norfolk	Set on fire and destroyed	Builder, xxx. 423
1872. June	St. Martin's Church, Birmingham	Slightly damaged	Builder, xxx. 423
1872. May	Rainham Church, Kent	Damaged	Builder, xxx. 391
1872. May	Mashbury Church, Essex	Set on fire	Builder, xxx. 391
1872. May	Bampton Parish Church	Much damaged	Builder, xxx. 391
1872. June	Chiddingley Church	Slightly damaged	Builder, xxx. 484
1872. June	All Saints' School, Little Horton	Slightly damaged	Builder, xxx. 484
1872. June	Kibblesworth Wesleyan Chapel	Slightly damaged	Builder, xxx. 484
1872. July	Brixton Church	Considerably damaged	Builder, xxx. 603
1872. July	Leigh Church	Severely injured	Builder, xxx. 591
1872. Aug.	St. Giles, Cripplegate	Slightly damaged	Builder, xxx. 629
1872. Aug.	Holy Trinity Church, Windsor	Severely injured	Builder, xxx. 610
1872. Sept.	Dundonald Parish Church	Spire and roof damaged	
1873. Apl.	Parish Church, Cromer	Slightly damaged	Builder, xxxi. 331
1873. Apl.	Martham Church	Much damaged	Builder, xxxi. 331
1873. Nov.	Ripponden Church	Much damaged	Builder, xxxi. 875
1873. Nov.	Industrial School, Mosbank, Glasgow	Set on fire	Builder, xxxi. 875
1874. July	Chesterfield Church	Slightly damaged	Builder, xxxii. 613
1874. July	Christ Church, Salford	Slightly damaged	Builder, xxxii. 613
1874. July	St. Luke's, Homerton	Set on fire, much damaged	Builder, xxxii. 613
1874. July	General Post Office, St. Martin's le Grand	Slightly damaged	Builder, xxxii. 613
1874. July	Military Prison, R.A. Barracks, Woolwich	Slightly damaged	Builder, xxxii. 613
1874. July	Free Church of Braco, Perthshire	Completely destroyed	Builder, xxxii. 613
1874. July	Ayot St. Peter Parish Church, Herts	Completely destroyed	Ann. Reg. p. 70
1875. June	Chester le Street, Durham	Spire considerably damaged	Newcastle Chronicle, June 16th
1875. June	West End Church, near Southampton	Spire destroyed	Builder, xxxiii. 586
1875. June	London and South Western Railway Co.'s tall chimney shaft at Southampton	Destroyed	Builder, xxxiii. 586

PUBLIC BUILDINGS STRUCK BY LIGHTNING (*continued*).

Date	Building	Damage	Authority
1875. July	Barthomley Church, near Crewe	Damaged	Daily paper
1875. July	St. Mary's Church, Birkenhead	Much damaged	Builder, xxxiii. 632
1875. Aug.	St. Nicholas Church, Blundellsands	Much damaged	Builder, xxxiii. 783
1876. Mar.	Cottingham Church, near Hull	Set on fire	Daily paper
1876. April	Snettisham Church	Considerably damaged	Daily paper
1876. April	Shotts Parish Church	Steeple destroyed	Daily paper
1876. July	Union Workhouse, Retford	Roof set on fire	Daily Chronicle, July 25
1876. July	Bishopstone Church	Considerably damaged	Lloyd's Weekly News, July 23
1876. July	Wilmcote Church	Considerably damaged	Lloyd's Weekly News, July 23
1876. July	St. Peter's Church, Stratford-on-Avon	Considerably damaged	Sunday Times, July 23
1876. Sept.	Grey Friars Tower, King's Lynn	Considerably damaged	Daily paper
1876. July	Market Hall, Doncaster	Damaged	Daily Telegraph, July 24
1877. May	Catholic Church, Wieschen, Poland	Six persons killed and seventy seriously injured	Globe, May 31, 1877
1877. May	All Saints Church, Stand Whiteland, Lancashire	Much damaged	Builder's Weekly Reporter, May 25, 1877
1878. May	Sir David Baird's monument, Perthshire	Almost entirely destroyed	Daily Telegraph. May 30
1878. June	St. Luke's Church, Hackney	Damaged	Daily paper
1878. July	Wesleyan Chapel, Southampton	Damaged	Daily paper
1878. July	Free Methodist Church, Tamworth	Damaged	Daily paper
1878. July	St. Jude's Church, Bethnal Green	Much damaged	Daily paper
1878. July	Church of the Holy Nativity, Knowle	Considerably damaged	The Times, July 27
1879. April	Henlow Church, Bedfordshire	Considerably damaged	The Times, April 18
1879. May	Laughten-en-le-Morthen Church	Considerably damaged	The Times, May
1879. June	St. Marie's Church, Rugby	Set fire to the woodwork	Weekly Dispatch, June 8
1879. June	Clevedon Market House, nr. Bristol	Very much damaged	Daily Chronicle, June 10
1879. Aug.	Parish Church, Wells, Norfolk	Burnt to the ground	Norwich paper
1879. Aug.	Cromer Church	Pinnacle damaged	Daily paper
1879. Aug.	St. Bride's Church, Stepney	Slightly damaged	Daily paper
1879. Sept.	Sanctuary of Madonna de Valmala, Valmala	Damaged. Several persons killed	Electrician, Sept. 6

POWDER MAGAZINES STRUCK BY LIGHTNING.

Date	Building	Damage
1732. Oct.	Gunpowder Magazine at Compost Major, Portugal	Exploded. City laid in ruins; above 1,000 people injured
1739. Sept.	Bremen	1,000 houses destroyed
1763. Nov.	Fort Augusta, Jamaica, powder magazine, containing 2,850 barrels of powder	Great number killed; much damage to property
1769. Aug.	Brescia Magazine, containing 207,000 lbs. of powder	Exploded; 3,000 persons killed.
1769	Venice	400 persons killed
1772. Nov.	Chester	Great damage to property; many lives lost
1773	Cambray	18 people killed; several houses greatly damaged
1773	Abbeville	150 persons killed; 100 houses destroyed
1780. Aug.	Malaga Gunpowder Magazine	
1782. Mar.	Sumatra Gunpowder Magazine	
1785. May	Tangiers Gunpowder Magazine	
1807. June	Luxembourg Gunpowder Magazine	About 12 tons of powder exploded
1808. Sept.	Venice Gunpowder Magazine	
1829. Nov.	Navarino Gunpowder Magazine	17 killed; 78 wounded
1840. June	Bombay Gunpowder Works	
1843. April	Dum Dum Gunpowder Magazine Sicily, Puzzaloni Gunpowder Magazine	
1843. April	Spain, Gaucin Gunpowder Magazine	A number of persons killed; church and 200 houses destroyed
1853	Hounslow Gunpowder Magazine	
1855. Oct.	Firework manufactory, Liverpool	Exploded
1856. Nov.	Rhodes Gunpowder Magazine	A considerable number of persons killed, and a large portion of the town laid in ruins
1857. Aug.	Bombay, Joudpore	About 1,000 persons killed; 500 houses destroyed
1878. Aug.	Bruntcliffe Colliery, near Leeds; powder magazine, containing about one ton of powder	Exploded
1878. Aug.	Pottsville, Pa. U.S.; a powder magazine containing 25,000 lbs. of powder	Exploded; 3 persons killed, several injured; many houses wrecked

CHAPTER XV.

THE EARTH CONNECTION.

To dwell too largely upon the importance of leading all lightning conductors down into moist earth, or, as technically called, 'good earth,' would be scarcely possible. It would perhaps not be too strong an expression to say that the part of the conductor above ground is a mere appendage to that under ground, the essential function of the whole apparatus—that of dispersing the electric force harmlessly —being accomplished by the subterranean portion. The clear understanding of Benjamin Franklin perceived this at the outset; but after him it seemed as if forgotten for a long time, and the result showed itself in numerous disasters that occurred to buildings protected with conductors, which brought the latter into disrepute with many persons. While, no doubt, in many instances the cause of these disasters was in the bad application of the conductors themselves, their defective character, or their feebleness, still in the great majority the underground connection may be taken to have been in fault. It may be laid down as an absolute certainty that a really good conductor—say, a copper rope from five-eighths to three-quarters of an inch in thickness—cannot possibly fail to carry off the electric force if the lower part reaches moist earth or water. Probably, in nine cases out of ten, whenever a building provided with a conductor is struck by lightning, it is for want of 'good earth.'

Franklin's own ideas were very clear on the subject. He laid them down at various times, more particularly when residing in England, during the years from 1764 to 1775, as colonial agent for Pennsylvania. During the latter part of this period he took an active interest in the proceedings of the Royal Society; and this learned body being requested by the Government to give advice regarding the best protection against lightning that could be provided for the great powder magazines at Purfleet, he was nominated into a committee with three other members, William Watson, H. Cavendish, and J. Robertson. The committee drew up a report, dated August 21, 1772, signed by all the members, but known to be written by Franklin alone. Dwelling strongly on the importance of the underground connection, Franklin says in this report: 'In common cases it has been judged sufficient if the lower parts of the conductor were sunk three or four feet into the ground, till it came to moist earth; but this being a case of great consequence, we are of opinion that greater precaution should be taken. Therefore we would advise that at each end of each magazine a well should be dug, in or through the chalk, so deep as to have in it at least four feet of standing water. From the bottom of this water should rise a piece of leaden pipe to, or near, the surface of the ground, where it should be joined to the end of an upright bar.' Franklin then goes on to recommend the usefulness of having even more wells than the two, so as to avoid any possibility of failure in protecting the powder magazines. 'We also advise,' he says in his report, 'in consideration of the great length of the buildings, that two wells of the same depth with the others should be dug within twelve feet of the doors of the two outside magazines —that is to say, one of them on the north side of the north building, and the other on the south side of the south building, from the bottom of which wells similar conductors should be carried up.' It is not on record whether these recommendations were adopted by the Government, but it seems likely that this was the case, as the fear of explosion of powder magazines through a stroke of lightning was very great at the time. Not long before, a magazine had been so destroyed at Brescia, in Italy, with the appalling result of a considerable part of the city being laid in ruins, burying many hundreds of persons. The destruction of the Brescia powder magazine, like all similar events, had, it is scarcely necessary to say, its due effect in spreading a desire for lightning conductors, fear doing what was not effected by foresight.

Whether or not the English Government made the wells recommended by Franklin for the Purfleet powder magazine, it is certain that the sound advice given was not largely followed. On the contrary, there grew a generally prevailing laxity in regard to the indispensableness of a good underground connection, which led to numerous accidents. They were seldom, however, ascribed to the right cause, others being sought instead—such as particular forms of conductors and the insufficient length of those phantoms called 'reception-rods,' which, as many thought, could never be made high enough, in order to 'draw the electric fluid' from the clouds. Height was sought where nothing but depth was required, and the same unsightly rods, towering high above buildings, would have very effectually carried off the electric forces if brought from the top to the bottom of the conductor, being taken out of the air and stuck into the earth. Still, there were not wanting philosophical minds impressed with the truth that no lightning conductor can discharge its functions unless rooted in moisture, and who not only knew it, but did their best to spread this knowledge in all directions. One of these philosophers, a singular character in his way, was a German clergyman, the Rev. Dr. Hemmer, who lived at Mannheim, on the Rhine, at the end of the last century. Taking the deepest interest in Franklin's great discovery, he made many experiments with lightning conductors, which brought him to the conviction that the electric force, in its chief tendency, seeks the mass of water on the globe, and that where this is not on the surface, it must be guided to it to become harmless. Consequently, he recommended to sink the conductor invariably deep into the ground, so as to reach water, and to subordinate everything else to this prime necessity. To make the use of lightning conductors as general as possible, Dr. Hemmer not only wrote a number of little books, which he liberally distributed, but travelled about through many parts of Germany, instigating the authorities to place conductors on all public buildings, and the people to set them up over their own houses. Holding that the earth connection was everything, he advocated simply to dig a hole in the ground till water or very moist earth was reached, and to stick a small iron bar, wrapped in lead to prevent rust, into it, running up the roof. The bar any village blacksmith could forge, and the hole any man or boy could dig, thus making the absolute cost of the conductor under this arrangement very trifling. Dr. Hemmer was right, no doubt, in his main argument, and most successful in spreading the knowledge of lightning conductors, while he was able to boast that not one of all the number he had set up had ever failed. However, he lived in an age when as yet water and gas pipes were unknown, and iron, or any other metal, scarcely entered into the construction of buildings. Given a leaden roof and a network of metal tubes, and Dr. Hemmer's small iron rod could scarcely be expected to do its work of protection.

Together with Dr. Hemmer in Germany, Professor Landriani, of Milan, drew attention to the paramount importance of a perfect earth connection. He made it his special business to investigate cases in which buildings with lightning conductors had been struck, and was able to show in nearly every instance that it had been for want of 'good earth.' A very striking case, which ought to have brought conviction of the truth to all investigators of the subject, occurred in Genoa in 1779. The church of St. Mary in this city, standing in a very elevated position, had been frequently struck by lightning, sometimes as often as twice in one year, and it was noticed that the electric force always followed precisely the same path, running along a certain portion of masonry, partly secured by iron hoops, and finally demolishing a wall at the bottom to get into the earth. At last, in November 1778, a conductor, made of the most approved design, was placed over the church, but, to the great surprise of the scientific men who had superintended the work, the lightning fell once more upon the building in the month of July of the following year, again following the old path it had constantly taken before, and causing absolutely the same damage as previously, even to the knocking out of certain portions of the wall nearest the ground. Naturally, the event caused widespread interest, leading to the closest examination of the church of St. Mary by several experts, among them Professor Landriani. He had no great trouble in discovering both the causes of the path of the lightning having always been the same when falling upon the church, and of the edifice having been struck again in the same manner when provided with a lightning conductor. Being a somewhat peculiar structure, consisting in part of hewn stones held together with iron cramps, there was a large quantity of metal both in and outside; and it was found that the path of the lightning had always been precisely in the direction where the metal offered the greatest continuity, leaping over the short intervals that existed by destroying the stone, and finally getting into the ground to a place where there was always a collection of water by knocking down a wall. If this accounted satisfactorily for the former accidents, that which took place when a conductor had been placed was not much more difficult of explanation. Professor Landriani found that though the conductor itself was very good, it was useless simply by having its roots in hard rock instead of moist ground. On the one side of St. Mary's Church there was a rill of water rippling down from the hills, and forming a small pool near the church, while on the other was the hard rock. It was into a crevice of the latter that the conductor had been laid, thus leaving the electric force to seek its old path into the water along the iron bars which, although disjointed, formed a far better road to earth than the planned road. It was a convincing proof of the supreme necessity of a good earth connection. Still, a long time yet was to elapse before conviction became general.

Probably, the matter was more studied by Italian scientific men than any others, the study of electricity having always been a favourite pursuit in that country; yet there, too, the matter was not understood till quite recently. This is proved by a letter of the celebrated astronomer and meteorologist, Father Secchi, addressed to the French scientific journal 'Les Mondes,' in October 1872, in which he tells the story of an accident that befel a building protected by lightning conductors set up under his own direction, the earth connection being made after rules laid down by Professor Matteucci, considered the leading authority on the subject. The letter of Father Secchi, though of some length, is given here entirely, both on account of the great fame of the writer, but recently deceased, and because it throws a flood of light on some of the most important points connected with the art of designing and applying lightning conductors.

'Eight years ago,' says Father Secchi, writing, as just mentioned, in 1872, 'some lightning conductors had been erected under my direction on the cathedral and on the Bishop's palace of Alatri, situated at the summit of the Acropolis of that town, which, by its elevated and solitary position, was exposed to frequent ravages from storms. It was not long ago that a flash of lightning demolished a great part of the belfry, and damaged the organ of the church. In the erection of this lightning conductor there arose a great difficulty proceeding from the nature of the soil, which at the depth of some centimetres turns out to be entirely of solid calcareous rock.

'In order to remedy this defect, that part of the conductor which enters the ground has been made very long, more than 4 metres [13 feet], and has been provided with a great many couples of points, 5 centimetres [2 inches] broad, 5 millimetres [⅛ inch] thick, indentated on the edges, with the addition of a thick copper wire twisted among the same points, to help to multiply the points of contact between the rod and the carbon. The foot of the lightning conductor is entirely of copper. The rod is also of copper up to a metre [3¼ feet] above the ground; and there is joined to it the iron conductor, in the ordinary receptacle made in the heart of the wall, to preserve it from disturbances of the inferior parts. The ditch into which the foot of the lightning conductor was sunk is 5 metres [16′ feet] long, and half-a-metre [1⅜ feet] wide, and it was dug into the ground as far as to touch the roots of some neighbouring trees, from which point upwards a layer of cinders was placed, covering the greater part of the ditch. Thus the surface of contact between the metal and the carbon, and of the latter with the soil, was such that one would have supposed it to be more than sufficient, while the presence of trees, although they were not very large, made it highly probable that the ground did always contain sufficient moisture. Moreover, as the edifice had two culminating points—namely, the belfry and the raised back portion of the choir—two rods were placed on them, each having an independent connection with the earth, so that, in the case of a discharge on one of the points, the electric force might find two ways in its course towards the earth.

'These arrangements produced, on the whole, a good result, since, although the edifice was struck at least four times after conductors had been placed on it, it suffered no damage of any kind. Nevertheless a very curious accident, highly interesting as a scientific study, happened on October 2. Early on the morning of this day several flashes of lightning fell down from the clouds during a terrific storm, which lasted over two hours. The belfry was struck at first by weak discharges twice; but the third flash was so appalling in its strength as to terrify the whole town below. The injuries it caused were not great, still they seemed to me to be extremely noteworthy. But before I describe them I must give some necessary details as to place and position of the lightning conductor.

'It so happened that four years after the erection of the conductor a line of pipes was laid down to carry water to the towns of Alatri and Ferentino, passing at a short distance from the belfry of the cathedral. The lightning conductor was not placed in communication with the pipes, because it seemed established, from previous experiments and observations, that it was needless to do so, the ground containing apparently sufficient moisture, the head of the waterworks being close, and there existing also a running fountain. I was not asked at the time whether it was necessary to establish this communication, but, had the question been put to me, I should probably have answered it in the negative, considering, from what I then knew, the work as superfluous. That I was in error then as to the necessities of a perfect underground connection is shown by what happened during the great storm in the early morning of October 2. The heavy flash of lightning before referred to did not go its appointed path underground, but passed off into the waterworks, with the following results:—

'1. It made in the earth a perfectly rectilinear excavation, which, from the lower part of the conductor, went to the tube of the waterworks running to Ferentino, and in traversing the wall destroyed the angle of that structure. The earth of the ditch thus dug was disposed regularly to right and left with great symmetry. The length of the ditch was about 10 metres, the depth about 70 centimetres [28 inches].

'2. The lightning struck the water-pipe of Ferentino, broke it completely, throwing the pieces to a distance of about 80 centimetres [32 inches]. The lead which soldered the joint of the broken tube with the tube beyond was found melted. In consequence of this rupture the water ceased flowing to Ferentino, and poured into the waterworks.

'3. Another part of the discharge spread itself by the pipe which goes to Alatri, and traversing the reservoir threw to a great distance some wooden plugs which stopped up the discharging tubes, the plugs being forcibly hammered in. It arrived at the town in a tank, where it damaged and twisted in a strange manner a leaden slab which was in the tank, made some other little injuries, and finally left the trace of its passage at the spouts of the public fountain.

'4. The point of the lightning conductor was examined, and it was found very blunt; it was found impossible to unscrew it, and it could not be removed without breaking the screw. It was found broken to a length of more than 3 centimetres [1¼ inch], and the section of fusion was nearly flat, as though it had been cut. The gold of the gilding had nearly all disappeared. In the church, and in the edifice which is attached to it, no injury was detected. These facts appear to me important both as regards practice and theory: in respect to theory, because they give an idea of the quantity and of the immense force of the discharge. The melting of the point down to a section 1 centimetre [½ inch] in diameter proves that it would have been melted down much further if it had been slighter. It is not prudent, then, to use very slender points; it is best that they should thicken quickly.

'The excavation of the ditch at the foot of the lightning conductor could not be the direct effect of electricity, but would be the result of the sudden evaporation of the moisture of the ground, generating steam, and forming, as it were, a mine.

'The breaking of the tube is most singular. It seems to me that it can with difficulty be attributed to the mechanical shock of the electricity itself. As the lead which united the broken tube to the one beyond was found melted, it is evident that, in spite of the water which flowed in this tube, it was raised to an enormous temperature in the place where it was struck, and probably it was the instantaneous evaporation of the water inside which caused the breaking of the tube.

'But the most singular fact, in a certain respect, is what was observed in the tube which descends to Alatri—that is to say, the alteration in form of the leaden slab. The little interruption which necessarily exists in this tank between the conducting-pipe and the metallic receptacle evidently gave occasion for a discharge by a flash, and, in consequence, for an explosion of steam. But we see at the same time by that that the distance traversed in the tube from the building to the slab, a distance of more than 200 metres [650 feet], in which the pipe is buried underground, did not suffice for the charge to lose itself in the ground, although during the passage it had to cross the reservoir, and might there have distributed itself. Our surprise is still greater when we reflect that it was only part of the discharge, since the greater portion had to flow by the water-pipe of Ferentino, which was the first struck in a direct manner, and that these pipes are joined together with lead. The quantity of electricity must have been enormous, in order to be able to have so much force and to run another 300 metres [975 feet] to reach the public fountain, and leave its traces there. A circumstance which deserves attention is, that this storm took place after a long and constant drought; and consequently the earth was less moist, and could offer little facility for dispersion.

'These cases are not so rare among us as one might suppose. Not very long ago, at Lavinia, a flash of lightning destroyed a great part of the belfry, passed to the bell, broke and melted it in its passage in such a manner that the metal had run away like wax. I do not believe this breakage of the bell to have been a mechanical effect of the lightning in a rigorous sense, for the bell could have been broken by the instantaneous expansion produced by the heat at the point of the passage, an expansion which had had no time to disperse, as a glass vase breaks when touched with a red-hot iron.

'Let these facts come about how they may, they enable us to see that it is necessary to devote great attention in the erection of lightning conductors, that we must allow them a large surface for discharge, *and that there can never be too much of it*. The surface of the foot of our lightning conductor was certainly superior to what has been judged sufficient by Matteucci for the discharges of telegraphic conductors, and yet it has not sufficed. Further, it is a confirmation of the necessity of making the neighbouring metallic masses communicate, and especially with water and gas-pipes.'

From out of the almost endless number of cases in which lightning conductors failed for want of a good earth connection, another one or two may be given, illustrated as having happened quite recently in England, and as such showing, in a very striking manner, in what a neglected state the knowledge of the subject still is at this moment. ‘A thunderstorm passed over the town of Clevedon, Somersetshire, in the afternoon of March 15, 1876, and a flash of lightning fell upon the steeple of Christchurch, provided, as was generally thought, with a most efficient conductor of recent construction, made of good copper rope. What happened is graphically and minutely told in a letter addressed to the 'Journal of the Society of Telegraph Engineers,' by Mr. Eustace Buttor, of Lewesfell, near Clevedon. 'There was but a single flash,' Mr. Buttor relates, ' which appeared to many observers to travel horizontally through the air. However, the lightning passed down the lightning conductor of Christchurch. The flag-staff, about 100 feet high, and the four pinnacles, about 90 feet high, have each a conductor, the flag-staff having the usual conical point, the pinnacles having the copper rope attached to their vanes. The five copper ropes unite inside the tower in the neighbourhood of the clock. Lower down the conductor passes through a slanting hole to the outside, and for the lowest 12 feet is encased in a pipe. On reaching the ground it passes into a dry freestone channel for about a dozen feet, and then dips down into the drain which carries rain-water from the roof. As no rain preceded or accompanied the flash, it may be presumed that *the drain was dry*.

'The protector is copper throughout, and, with the exception of the termination, seems to have been carefully and efficiently placed. The diameter I estimate to be half-inch, or it may be a trifle more. Just at the point where it leaves the pipe and enters the ground, the electric charge left it, dashed through three feet or more of solid wall supporting the tower, in order to reach the gas-meter inside, then it passed safely along the gas-pipe. The cavity made was considerable, but very irregular. I was unable to ascertain when the workmen were engaged in repairs, and therefore cannot give their estimate of the weight of stone displaced, but it must have been many hundredweights, though only a few pounds were actually thrown out on to the path, or inside into the vault. A large quantity of stone was pulverised, and the whole gave one the idea of the explosion of a charge of gunpowder under great compression. In a house about 100 yards from the church, the inmates felt the shock intensely, but did not know that the house had been touched. Some hours after, however, on going to turn on the gas, a hissing noise was heard, and a hole was found in the composition gas-pipe, about five-eighth inch diameter, just where the pipe passed within an inch of a water-pipe. The lightning must have come along the main from the church gas-pipe to this house, and then passed to the water-pipe as the readiest way to moist earth. The whole soil in the neighbourhood is mountain limestone, very dry. There is not the slightest evidence of displaced plaster, or any other sign of the passage of an electrical discharge through the house.' There need be little comment on the facts stated in this letter, notable though they are. It is the old delusion that a lightning conductor need be brought down underground only, and that then all is right. In this case, those who protected Christchurch, Clevedon, thought it quite sufficient to bring the conductor down into a drain-pipe carrying rain-water from the roof, without reflecting for a moment that an earthenware drain-pipe would insulate the conductor from 'earth.' A similar instance came under the writer's notice about a year ago. One of the pinnacles of Cromer Church, in Norfolk, was struck by lightning, although fitted with a conductor on one of the pinnacles. On examination it was discovered that the earth terminal had been inserted into an earthenware drain.

It is not very easy to give exact prescriptions as to the best manner in which the underground connection should be effected. The means vary entirely with the circumstances, and the matter should in all cases be intrusted to an expert. Simple as is the whole theory of lightning protection, consisting in nothing else but laying a good metallic path from the top of a building down into moist earth, as an unfailing path for the electric force, the practical execution of it is not the less often very complicated. It is especially so as regards the most important of points, that of the underground connection. Of course, wherever there is running water at hand, a river, or even a tiny stream that never dries, the matter is easy enough, but as in the great majority of buildings to be protected such water does not exist, the solution of the question becomes more difficult, and frequently one of the greatest perplexity. It tends even to be more and more so in consequence of the progress of sanitary arrangements under which towns and villages are 'drained' until the soil has been made as dry as a rock. Immense as the benefit is to public health, it is, like all benefits, attended by certain drawbacks. One of these certainly is a greater danger from lightning. It is often proposed by builders to use the drain-pipes themselves in making 'good earth' for lightning conductors, but the fallacy of this recommendation need scarcely be exposed, seeing that these conduits are generally made of earthenware, as happened when Christchurch, Clevedon, was struck by lightning.

While broad rules cannot be laid down, still it may be affirmed that a good earth connection, sufficient to carry off the heaven's electric discharges, may always be obtained by either of two means. The first, and in all cases most preferable, is to lay the conductor deep enough into the ground to reach permanent moisture. When this exists in a considerable mass, the single conducting rope, touching it, will be quite sufficient; but when the quantity is deficient, or doubtful, it will certainly be advisable to spread out the rope,

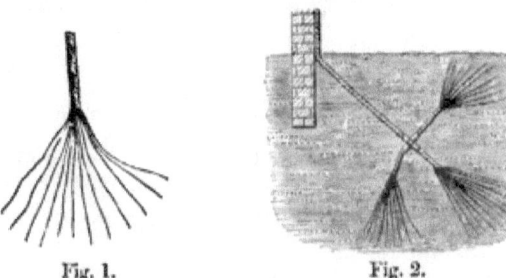

Fig. 1.   Fig. 2.

so as to run in various directions, similar to the root of a tree, likewise in search of moisture. There are various modes of accomplishing this, shown in figs. 1 and 2.

A variety of methods have been proposed for the dispersion of the electric force underground where the soil contains little or no moisture, except at great depths, to be

Fig. 3.

reached only by a vast amount of labour and expenditure. In France, the system most generally adopted in these cases is to place at the bottom of the underground connection an apparatus, made either of iron or copper, shaped somewhat in the form of a harrow, and to embed it thickly in charcoal. Fig. 3 will illustrate this system of earth connection.

The apparatus is as simple as it may be useful, and the more so, of course, the thicker the mass of charcoal in which it is embedded. But it may be doubted whether it is sufficient to make 'good earth' under all circumstances. Perhaps it will do so in ninety-nine cases and fail in the hundredth. The amount of electric force discharged in ordinary thunderstorms does not seem to vary much, and, according to all observations, such an artificial connection as this of the charcoal bed is sufficient to disperse it safely beneath the surface. But now and then there come storms of extraordinary violence, or, in other words, extraordinary accumulations of atmosphere electricity, which demand precautions such as are not fulfilled by the subterranean harrow, however thickly embedded in charcoal, or, as oftener done, in gas coke or cinders. It is certain that there have been cases in which buildings with otherwise excellent conductors, but provided with such an artificial earth connection, have been damaged by lightning. However, it may be stated, as the net result of all observations and known facts upon the subject, that small private houses can be well protected by this means against lightning, but that the system cannot be recommended as absolutely safe for large edifices and public buildings.

To protect any structure of great extent, it is absolutely necessary to bring the conductor, or conductors, deep enough into the earth to reach water. It is all the more indispensable with modern buildings, as they contain large masses of metal, not only in gas and water-pipes, but often in staircases and iron columns, towards which the electric force has the strongest tendency to direct itself unless drawn to the earth by an immediate and unfailing connection with the great sheet of water below its surface. It is considered by German electricians that there is no necessity, if a large edifice has a number of conductors, to let each have a separate earth connection; it is quite sufficient to bring them all into one, provided only that this is absolutely perfect at all seasons

THE EARTH CONNECTION. 213

and under all circumstances. Fig. 4 will show how this can be done.

It will be seen that for the protection of this edifice there are six conductors, with four elevated points marked A, B, C and c. Two of these points, A and c, expand from the roof

Fig. 4.

to the ground into double conductors, so as to protect the sides of the building against possible lateral discharges of lightning, and all the six conductors meet a little below the surface in the earth connection prepared for them. To form this one connection, either by digging or boring, may sometimes be costly, but whether the expenses be more or less, the protection against lightning thus effected will be so absolute as to be invaluable.

In a similar manner as the large edifice, with its many

Fig. 5.

gables, a church may be fitted with lightning conductors. Fig. 5 scarcely needs much explanation.

There is one thing, however, regarding churches, that must be well borne in mind in establishing their protection against lightning. Besides containing great masses of metal, in bells, organs, and other contents, they are frequently placed in high situations, exposed to the most violent discharges of the electric force. It often happens also that they stand on rocky ground, with the subterranean waters far below the surface. To ensure absolute protection under these circumstances, it is indispensable to connect the conductors with water, wherever it is to be found, by a solid channel, into which the copper rods may run, if possibly some distance below the surface of the earth. The form such a channel may take is indicated on the engraving. It will be seen that the protection against lightning indicated here is not only for the church, but the adjoining parsonage, the conductors spreading over both, with points on the most prominent and exposed places. It would be possible to carry out this principle in ensuring the protection of a whole block of private buildings. German electricians think that one channel or well, sufficiently broad, leading from the surface of the earth to layers always moist, or to perennial springs, would suffice to carry the electric force discharged upon a hundred conductors, and all the easier as it would be impossible that many would be struck at one and the same time by lightning. Perhaps some such arrangements will be made in the future, when both houses and towns are built upon a more systematic plan than is followed at the present time.

If, as a rule, one channel of underground connection is amply sufficient for the protection of even the largest buildings, there may be cases in which it is indispensable to spread the conductors into several directions. It may be laid down, broadly, that when there is water to be reached, the one channel is sufficient, but that when this is not possible, or expedient, more lines of underground connection must be formed. Fig. 6 may serve to illustrate a case of the latter kind. It shows a powder-magazine, partly above and partly underground, standing on dry soil, with trees in the neighbourhood, likely to add to the danger of atmospheric discharges of electricity, and with no stream, or permanent moisture, into which to guide them. Nothing remains, under these circumstances, to ensure safety, but to multiply the lines of underground connection to the utmost extent. To add to the facility of the dispersion of the electric force, the main channels may be filled with charcoal, broken coke, or cinders, and if large quantities of these substances can be placed in one or two pits, it is possible to make thus an artificial connection as nearly as can be responding to 'good earth.' Still, it must never be forgotten that, absolutely, 'good earth' in reference to lightning conductors means moisture, or water.

If permanent moisture cannot be obtained and iron

Fig. 6.

water-mains are within reach, it is desirable to connect the ground terminal with them by means of good solder, as from the large mass of metal they generally form very good 'earths.'

In giving directions, or rather suggestions, about the design and application of conductors, and, what is most important in regard to them, their connection with the subterranean mass of waters, the idea that persons may construct their own conductors is left aside altogether as absurd. It is a good old proverb which says that a man who is his own lawyer is certain to lose his cause; another has it that a man who is his own doctor is sure to succumb to his illness. With regard to the setting-up of lightning conductors, it is precisely the same. Simple enough as is the theory of 'drawing lightning' from the clouds, the practical execution of it is, as mentioned more than once, not a little complicated. The formation of the underground connection, in particular, is a matter requiring very great experience, and very frequently one of the utmost difficulty. Vast sums of money are often thrown away needlessly in making a connection which in the end proves useless, while, on the other hand, a trifling addition to the expenditure in setting-up a conductor would procure its efficiency, not attained simply from want of 'good earth.' A recent writer on lightning conductors whimsically, yet with much truth, expresses it by remarking that 'people spend money upon gilded points on the top of the house, while they ought rather to sink it in water at the bottom.' Undoubtedly, the efficiency of conductors lies, even more than at the top, on 'the bottom.' The earth connection may be called 'the alpha and omega' of lightning protection.

CHAPTER XVI

INSPECTION OF LIGHTNING CONDUCTORS.

THERE is one subject in regard to the proper protection of buildings against the destructive effects of lightning which is generally overlooked, at least in this country, to a really surprising degree. It is the necessity that lightning conductors, once put up, should be regularly inspected, to see if they are in good order, so as to be really efficacious. That this is very rarely done, is one of the main reasons why accidents by lightning sometimes occur in places nominally protected by conductors. The neglect is the more astounding, as one would think that all intelligent persons, whose knowledge prompted them to see the wisdom of protection against lightning, would likewise come to the conclusion that the scientific apparatus set up to effect it required occasional repairs, such as the clocks in their houses and the buildings themselves. But such is very far from being the case. It is, perhaps, not too much to assert that at present not one in a thousand persons who have gone to the expense of protecting their houses by lightning conductors make the protection complete, at a merely nominal cost, by providing a regular—say, annual or bi-annual—inspection.

The causes which necessitate such inspection are numerous. In the first instance, there is the constantly acting influence of wind and weather upon those parts of the conductor which are above earth. Wonderful as is the simple machinery devised by Franklin which conducts the mysterious electric force from the clouds into the ground, depriving it of its destructive power, it is, after all, but a feeble thing in itself, and necessarily so. The upper terminal of the conductor—what the Germans call the 'reception rod,' and the French the 'tige,' or stem—cannot be very thick without becoming unsightly, and, as regards large public buildings, destroying their architectural effects; while the rope, or ribbon, running to the ground must, for the same reason, as well as that of cost, be of comparatively small diameter. Subject to the constant effects of moisture, to wind, and ice, and hailstorm, there is always a possibility of the slender metal strips being damaged, so as to interrupt their continuity, and thus destroy the free passage of the electric force. Instances have happened in which the damage done was so slight as to be scarcely visible, and still sufficient to destroy the efficacy of the conductor. Nothing but the regular testing by a galvanometer—one of which is described, with an illustration, on page 60—by an experienced person can establish the fact that the action of the conductor remains perfect.

A second important cause for inspection lies in the necessity of always ascertaining with accuracy whether the earth connection is really in a faultless state. The immense significance of the earth connection—the basis, in more than one sense, of lightning protection—having been dwelt upon in the preceding chapter, it is only necessary here to state that, even if perfectly secured at the outset, it is liable to disarrangements. One not infrequent accident causing them is a change in the soil from moisture to dryness, which may be brought about either by altered drainage or long absence of rain. The dangers which threaten a break in the earth connection by altered or improved drainage are of the most serious kind, and likely to become more so from year to year. Not only the soil of our towns and cities, but even that of our villages, and the fields themselves, is getting ever more honeycombed by drain-pipes, until almost every drop of moisture is sucked out of the ground. No doubt the pipes themselves may improve the earth connection, if of iron or any other metal. But very frequently they are of earthen ware, in which case they are far more dangerous than useful, even if filled with water. To guard against the danger likely to arise from changes in the drainage, it would be wise to have a thorough examination, by means of the test galvanometer, of all lightning conductors near to or affected by alterations in the drains, whenever completed. The same recommendation may be made as regards cases where the soil has become unusually dry after a long absence of rain. Few persons, except those who have made a study of the subject, can form an idea to what depth such dryness often extends, more especially in sandy and gravelly soils.

There is a third ground, as material for consideration as each of the two preceding ones, upon which the regular inspection of lightning conductors must be strongly urged. It is, that constant alterations in the interior of buildings, private residences as well as public edifices, may serve to destroy the efficacy of a conductor which was originally good, even to perfection. Thus a roof may be repaired, and lead or iron introduced where it was not before; or clamps of iron may be inserted in the walls of houses, to give them greater strength; or, in fact, any changes may be made which bring masses of metal more or less in proximity to the conductor. Under such circumstances, the efficacy of the conductor is destroyed just in proportion as the metal forms a better path for the dispersion of the electric force than the one artificially prepared. There are hundreds of instances to prove that changes made in buildings, such as the addition of a leaden roof without, or the iron balustrade of a staircase within, diverted the current of the electric force from the conductor on its way to the earth, originally well provided for. In one rather curious case, which happened at Lyons not many years ago, even an alteration of the fixtures of a house proved destructive to the efficacy of a conductor, perfect at the outset, the latter fact being shown in that it had previously received a stroke of lightning and brought it harmlessly to earth. The case was that of a banker possessed of the piece of furniture indispensable to his profession, namely, a large iron safe. It stood at first near an inner wall, in the centre of the house; but wishing to add to its strength in resisting the attack of burglars, the banker had it embedded partly in another wall adjoining that on the outside, near a place where the masonry was held together by some large iron clamps. In delightful ignorance of the effect of this removal of his safe inside the house upon the lightning conductor outside—an ignorance which would have been the same, probably, among 999 persons out of 1,000—the banker sat quietly down to dinner with his family one day in July, when a terrific shock made the whole house tremble to its foundations, upsetting furniture and breaking glasses. The idea of an earthquake naturally came up at once; but when looking out of the window (shivered to pieces) the banker was told by a crowd assembled outside that there had been no earthquake, but that his house had simply been struck by lightning, as it had been before. But while previously the electric force had passed silently into the ground, unknown even to the inmates of the house, and its passage verified only by the accidental observation of a neighbouring meteorologist, it had this time left its appointed path, seeking a new road more strongly attractive. The lightning had found its way into the banker's safe, filled with gold. Once inside, the electric current, not finding a farther outlet, had expended its force in shattering the walls and making the house tremble, besides melting some gold and burning banknotes. The investigation of the case at the time made some noise, but it had one most useful result—it led to the institution of a new office in connection with the Department of Public Architecture of the city of Lyons, that of an inspector of lightning conductors. He was charged to examine at stated intervals, or as often as circumstances seemed to require it, the conductors applied to all the public buildings of the city, to ascertain their efficacy, and, if not deeming them in good condition, to effect all necessary repairs. Shall we repeat, again and again, 'They manage things better in France'?

The regular inspection of lightning conductors, as yet unknown or all but unknown in England, has been for a long time in practice in several States of Continental Europe, among them Germany and France. The origin of such inspection may be traced to Northern Germany. It has been mentioned before (Chap. IV., page 43) that the first lightning conductor set up over a public building in Europe was erected on the steeple of the Church of St. Jacob, Hamburg, and that the extension of conductors in the city and neighbourhood was so rapid, that before five years had gone by there were over seven hundred conductors. 'To this day they are comparatively more numerous in this district than anywhere else in Europe.' To this day, too, the scientific aspect of the question of lightning protection, and the statistics connected with it, are more appreciated here, and have been more closely investigated, than in any other part of Europe. In recent years, this has been more particularly the case in the territories to the north of the city of Hamburg, the German province of Schleswig-Holstein. Not even in the country of their origin, and the one which, as yet, has the greatest number of them in use, have the 'Franklin rods' given rise to so much serious study as in that part of Germany.

Thunderstorms are more numerous, on the average, in Schleswig-Holstein than in any other part of Central and Northern Europe—due, probably, to the fact of the province not only being a narrow peninsula, with the Baltic on the east, and the German Ocean on the west, but intersected by rivers and canals, producing a generally moist atmosphere. Almost all public edifices in the province, and the great majority of private buildings above the rank of mere cottages, are protected by lightning conductors; and to aid in their extension there are special laws under which damages by lightning are not made good, except to a limited extent, by fire insurance companies, unless it is proved that the edifices struck had been provided previously with efficient conductors. These laws gave rise to a curious investigation some three or four years ago. It was found that the principal fire insurance office—an institution under the patronage of the Government, called the 'Landesbrandkasse,' or 'County Fire Insurance Office'—had been called upon a number of times to pay for damage caused by lightning in cases where the buildings were provided with lightning conductors of the best kind, in apparently perfect condition. Though the cases were very few indeed—namely, but four out of 552 claims for damages from lightning which had been made in the course of eight years—still, the interest taken in the subject was so great, that the managers of the institution appointed a special commissioner to inquire thoroughly into the matter as to how it could happen that buildings provided with proper conductors could ever be struck by lightning. The gentleman chosen to undertake this task was Dr. W. Holtz, of Greifswald, well known as having given much time to the study of the phenomena of electricity, as well as the construction of lightning conductors. Dr. Holtz in due course made his report, which was afterwards published in a scientific journal called 'Nachrichten des Naturwissenschaftlichen Vereins für Neuvorpommern und Rügen,' being the organ of a society under the latter title. The report—which must be completely unknown in this country—is full of interest, and well deserves being extracted from in several notable particulars.

Dr. Holtz begins his report by referring to the well-known fact, already dwelt upon, that in some instances lightning conductors have got into disrepute because houses provided with them have been struck and damaged. 'Unhappily,' he says, 'there are still at the present moment many persons who question the utility of conductors, simply because it happens now and then, that lightning, apparently in entire disregard of them, falls upon dwellings. These persons completely overlook two facts, namely: first, that such cases are excessively rare; and, secondly, what is far more important and more to the point, that it is beyond dispute that whenever buildings nominally provided with conductors are struck by lightning, these conductors are not in an efficient state. Such buildings are absolutely in the same condition as if they had no conductors at all.' Dr. Holtz then goes on to speak of his journey of inspection to inquire into the causes of failure, or so-called failure, of lightning conductors. He says that, having examined a vast number of conductors, he found that in a good many instances real use had been sacrificed to ornament. He expresses this somewhat quaintly, in scientific style, apparently with the intention of not giving offence to anybody—not even to manufacturers of lightning conductors. 'It was found by me,' Dr. Holtz states, 'that the unreal was frequently placed above the real, and that many lightning conductors, although very costly in the first instance, afforded no certain protection.' The meaning of this clearly is, that too much attention is given to the upper part of conductors, especially the pointed top—frequently covered with needless gilding—and far too little to the part underground, forming the all-important earth connection. It is a criticism true for other countries besides Germany.

Among the many interesting remarks of Dr. Holtz, evidently based on a thorough knowledge of the subject which he treats, are some good ones about the necessity of constructing lightning conductors, not slavishly after old models, but in conformity with modern requirements, carefully considering the nature of the buildings to be protected and their materials. 'The increase of metals,' he says, 'in the construction of houses, both inside and outwardly, is assuming larger proportions from year to year. An absolute consequence of it is, that the electric force called lightning is tempted, far more than was the case in older dwellings, not to go to the conductor at all, or, if attracted to it, to leave the path afterwards, seeking other attractions. I found this to have been the case, in the course of my investigations, in several instances, two of them notable ones.

The first was that of the public school of the town of Elmshorm, struck by lightning away from the conductor; and the second that of the church of St. Lawrence, in the town of Itzehoe, where the conductor was struck at first, but the lightning deviated subsequently from its metal path. In both cases I found that the non-efficacy of the conductor was caused by a number of gas-pipes. But there are many other metallic masses besides gas-pipes which interfere thus with the proper action of lightning conductors More or less, all metals do so, especially those which lead to the ground, or are in contact with moisture. Water-pipes will attract the electric force even more than gas-pipes, and likewise the metal tubes which carry the rain from the roof into the ground. But it may also happen that mere ornaments on the roof, more particularly if of thick metal, and carried all along the top and sides, may divert the electric force from the conductor, although they have no connection whatever with the ground. Even the many wires outside and inside houses, for bells and other purposes, may do mischief. There can be no doubt whatever that the large increase of the use of metals in the construction and ornamentation of modern houses has led to far greater danger to which they are exposed from lightning. At the same time there is equally little doubt that all this increased danger may be absolutely guarded against by the setting up of lightning conductors by competent persons, carefully designed to meet all cases.' Dr. Holtz adds, further on, that one most important element of protection to be obtained from conductors consists in the regular testing of them, without which, indeed, there can be no permanent security.

What the writer says on the inspection of conductors is particularly worth quoting. 'A lightning conductor,' he remarks, ' however excellent in the first instance, may lose all its good qualities, for several reasons. In the first instance it may suffer, like all mundane things, from age. The decrepitude will come on all the sooner whenever the materials are not of the best kind, or whenever little care has been taken in properly connecting the various parts. This is frequently the case in conductors of old design. But, even if all has been done that scientific skill can accomplish, age will make itself felt some time or other. Oxidation will play its part; so will the warfare of the elements. However safely secured at first, the attachment of the parts to the buildings will get loose, or perhaps even broken. Repairs consequently become indispensable. When are they to be effected? It can only be indicated by testing the conductor from time to time.' Dr. Holtz next dwells at some length on the necessity of conductors being designed by thoroughly competent persons; not mere 'lightning rod men,' who are able to take into account all the particulars of the building which is to be protected, more especially the metal employed in the construction. 'A conductor,' he truly remarks, 'cannot be expected to be a trustworthy protection against the destructive force of lightning, if simply set up over a house without consideration of its outer and inner features. Perhaps in buildings of olden times, into the construction of which metals seldom or never entered, a simple wire running from top to bottom, surmounted by an iron rod, was quite sufficient, but this is no longer the case, all the circumstances having been completely altered. The wire, however thin, was not merely the best, but the only path for the electric force. But at present the masses of metal used in the construction of buildings constitute a number of rival paths, and it requires very careful consideration indeed to lay down an absolutely infallible lightning conductor in such a way as to overcome all influences opposing its action. Therefore conductors of old construction can not only not be expected to be efficacious under modern exigencies, but even those made at the present time cannot be expected to be efficient under circumstances which, probably, the future may bring forth. There is really nothing else to make a lightning conductor a safe protection under all circumstances, and at all times, but regular, constant, and skilful examination.'

To the three great causes before indicated which make the regular testing of conductors an almost imperative necessity, several minor ones may be added. Among them may be cited the frequency of repairs of the walls and roofs of houses. Our modern houses, as we all know, are not built, like those of the Romans, 'for an eternity,' but in the vast majority, particularly in towns, are 'leaseholds for ninety-nine years.' Many of them, perhaps, can scarcely be expected to last ninety-nine years, being constructed by their builders on the principle of Peter Pindar's razors, 'not to shave, but to sell.' Hence the absolute necessity of repairs without end. Without casting the least slur upon the character of the artisans who execute these, bricklayers, plasterers, painters, plumbers, and others, it may be fairly asserted that they are densely ignorant as to the nature of lightning conductors. It is not at all a wonder that this should be so, since they share their ignorance with many persons of far higher education, who know no more of the action of the electric force in seeking its way from the clouds into moist earth than they do of that of a voltaic apparatus, or of a condensing steam-engine. These artisans, then, in whose hands the repairs of houses are left, naturally treat the narrow strip of metal running from the top of houses to the bottom with great indifference, not having the slightest idea of its being one of the most marvellous conceptions of the human mind. It has been reported, on good authority, that there are frequently workmen to be found, such as house painters and others, whose business it is to 'decorate' the outside of dwellings with the stuff called 'stucco,' who feel a sort of mild hatred for lightning conductors, as interfering with their achievements, and, as they think, disfiguring the beauties which they are creating. Woe to the poor conductor within their reach! Unless very conspicuously placed, which is rarely the case, the tenant of a house will seldom discover in time that the slender rope, or ribbon, which gives him and his family protection against lightning has been broken by cunning hands when the last repairs were effected, and the ends stowed away in the gutter on the roof. The discovery will be made, in the absence of inspection, probably, only under the fierce light of a flash of lightning from a passing thunderstorm.

If in towns the ever increasing accumulation of gas, of water, and of drainage pipes constitutes a danger against the efficacy of lightning conductors—to be guarded against only by frequent testing—there is another source of danger arising in the country. It is in the planting of new trees and the growth of old ones which is constantly going on in the vicinity of the thousands of country houses and mansions with which Great Britain is dotted from one end to the other, more than any other country in the world. The fact has already been dwelt upon, that trees are more liable to be struck by lightning than any other natural objects, the reason of it being unknown, except in the very probable surmise that the moisture in them forms the natural cause why the electric force seeks its path through them to the earth. Whatever the cause or causes, there can be no doubt that trees are incessantly struck by lightning, and that they are the more exposed to be struck the higher they are and the wider the extent of their branches. Consequently, wherever trees are being planted, or growing up around houses, the greatest care should be taken in designing lightning conductors, so as to provide against the action exercised by them in juxtaposition to the electric force. Thus, if trees, originally small, should reach to such a height above dwellings as to make it possible that a stroke of lightning will fall upon them, in preference to the conductor, the arrangements for protection will have to be altered, so as to ensure the safety of the house nearest these particular trees. Again, if, as often happens, there are new trees planted near a building the side of which has no protection whatever, such as a greenhouse or conservatory, the conductor should be extended in this direction. In connection with trees, mention must be made of wells and fountains, as possible dangers to the proper action of lightning conductors. Many a disaster has been caused by newly-made wells to dwellings which were previously well protected by conductors. The only safeguard against danger arising from these and numerous other causes, which it would be tedious to specify, lies in careful, constant inspection and testing of conductors.

It is lamentable to think that while the regular inspection of lightning conductors has been admitted long ago to be a necessity in many countries on the Continent of Europe, we as yet have taken no steps whatever to realise it. There is, probably, not a single public building in England which has conductors systematically tested from time to time. While there are tens of thousands of edifices, private and public, that are entirely without protection against lightning, there are many thousands of others which, nominally protected, are in reality in the same position. They have conductors, but it is impossible to say whether they would be efficacious were a more than usually heavy stroke of lightning to fall upon them. The inmates of such dwellings live in fancied security, which is the more to be deplored, as it would be so easy to make it real. All that is required is a knowledge of the subject. With the growth of such knowledge it is certain that the inspection of conductors will become general, with the good effect, above all others, of setting at rest all doubts as to the infallible security they afford, if properly constructed and maintained, against damage from lightning.

APPENDIX.

BIBLIOGRAPHY OF WORKS BEARING UPON LIGHTNING CONDUCTORS.

1663. HIER. CARDANI de Fulgure liber unus. Opera omnia. Lugd. tom. ii. pp. 720.
1666. DR. WALLIS. A Relation of an Accident by Thunder and Lightning at Oxford. Phil. Trans. i. 222.
    THOS. NEALE. Effect of Thunder and Lightning. Phil. Trans. i. 247.
1670. Effects of Lightning at Stralsund. Phil. Trans. v. 2084.
1676. On the Effects of Thunder and Lightning on Sea Compasses. Phil. Trans. xi. 647.
1683–4. DR. LISTER. On Thunder and Lightning. Phil. Trans. xiv. 512.
    SIR R. S. On the Effect of Thunder on the Compass of a Ship. Phil. Trans., xiv. 520.
1685. On some Remarkable Effects of a Great Storm of Thunder and Lightning at Portsmouth. Phil. Trans. xv. 1212.
1696. DR. GEO. GARDEN. On the Effects of a very Extraordinary Thunder-Storm near Aberdeen. Phil. Trans. xix. 311.
1697. DR. WALLIS. On Hail, Thunder, and Lightning. Phil. Trans. xix. 653.
    DR. WALLIS. On the Effects of Thunder and Lightning. Phil. Trans. xx. 5.
1708. S. MOLYNEUX. On the Effects of Thunder and Lightning. Phil. Trans. xxvi. 36.
    O. BRIDGMAN. On the Effects of Thunder and Lightning. Phil. Trans. xxvi. 137.
    WALL. Experiments on the Luminous Qualities of Amber, Diamonds, and Gum Lac. Phil. Trans. xxvi. 69.
    JOS. NELSON. On the Effects of Thunder and Lightning. Phil. Trans. xxvi. 140.
1709. R. THORESBY. On a Storm of Thunder and Lightning. Phil. Trans. xxvi. 280.
1725. REV. JOS. WASSE. On the Effects of Lightning. Phil. Trans. xxxiii. 366.
1730. EVAN DAVIES. On the Effects of Thunder, &c. Phil. Trans. xxxvi. 444.
1734. J. HENR. A. SEELEN. De Tonitru existentiae Dei teste. Miscellanea, P. I. 81. Lub.
1735. STEPHEN GRAY. On the Electrical Light. Phil. Trans. xxxix. 24.

1739. Sir Jno. Clark. On the Effects of Thunder on Trees. Phil. Trans. xli. 235.
1742. Lord Petrie. On the Effects of Lightning. Phil. Trans. xlii. 136.
1744. J. H. Winkler. Gedanken von den Eigenschaften, Würkungen und Ursachen der Elektricität. 8vo. Leipzig.
1745. J. H. Winkler. Die Eigenschaften der elektrischen Materie und des elektrischen Feuers aus verschiedenen neuen Versuchen erklärt. 8vo. Leipzig.
1746. J. H. Winkler. Abhandlung von dem elektrischen Ursprung des Wetterleuchtens.
J. H. Winkler. Von der Stärke der elektrischen Kraft des Wassers in gläsernen Röhren. 8vo. Leipzig.
1747. Maffei. Della Formazione dei Fulmini. 4to. Verona.
Wm. Watson. On the Velocity of Electricity. Phil. Trans. xlv. 49.
1748. Wm. Watson. Of the Experiments made by some Gentlemen of the Royal Society to measure the absolute Velocity of Electricity. Phil. Trans. xlv. 491.
1749. Abbé Nollet. Recherches sur les Causes particulières des Phénomènes électriques. 8vo. Paris.
1750. Barberet. Dissertation sur le Rapport qui existe entre les Phénomènes de Tonnerre et ceux de l'Electricité. 4to. Bordeaux.
1751. B. Franklin. Experiments and Observations in Electricity, made at Philadelphia, in America. 8vo. London.
B. Franklin. Concerning the Effects of Lightning. Phil. Trans. xlvii. 289.
Barberet. Discours, qui a remporté le Prix de Physique, au jugement de l'Acad. de Bordeaux, en 1750 : S'il y a quelques rapports entre les Phénomènes du Tonnerre et ceux de l'Electricité. 4to. Bordeaux.
A. G. Kästner. Nachricht von einer besonderen leuchtenden Erscheinung, so auf einem Thurme zu Nordhausen gesehen worden. Hamburger Magazin. vii. 420.
1752. Abbé Mazeas. On the Analogy of Lightning and Electricity. Phil. Trans. xlvii. 534.
B. Franklin. On the Electrical Kite. Phil. Trans. xlvii. 565.
H. Eeles. On the Cause of Thunder. Phil. Trans. xlvii. 524.
A. G. Kästner. Nachricht von einem besonderen Lichte. Hamb. Magaz. ix. 359.
J. G. Krull. Versuche zur Bestätigung der Meinung, dass die elektrische Materie mit der Materie des Donners und Blitzes eine grosse Aehnlichkeit habe. Hannover: Gelehrte Anzeigen vom J. 1752.
Le Monnier. Observations sur l'Electricité de l'Air. Mém de math. et de phys. de l'Acad. R. d. Sc. de Paris, A. 1752, p. 233. Paris. Biblioth. d. Sc. et d. Beaux Arts. vi. 38.
Ch. Mylius. Extract of a Letter from Mr. Mylius, of Berlin, to Mr. W. Watson, On extracting Electricity from the Clouds. Phil. Trans. xlvi. 559.
Ch. Mylius. Nachrichten und Gedanken von der Elektricität des Donners. Physik. Belustigungen. 8vo. p. 457. Berlin, 1752.
Abbé Nollet. Extracts of two Letters of the Abbé Nollet to Mr. W. Watson, On extracting Electricity from the Clouds. Phil. Trans. xlvii. 553.
W. Watson. Concerning the Electrical Experiments in England upon Thunder-Clouds. Phil. Trans. xlvii. 567.
1753. P. A. Bina. Elektr. Versuche, Gewitter und Regen betreffend. Hamburger Magaz. xii. 57.
J. Bunsen. Versuch, wie die Meteora des Donners und Blitzes, des

Aufsteigens der Dünste, incl. des Nordscheins, aus elektr. Versuchen. herzuleiten und zu erklären. 8vo. Lemgo.

1753. G. BECCARIA. Dell' Elettricismo artifiziale e naturale. 4to. Torino, 1753.
W. WATSON. On Nollet's Electricity. Phil. Trans. xlviii. 201.
M. MAZEAS. Electricity of the Air. Phil. Trans. xlviii. 377.
M. LOMONOSOW. Oratio de Meteoris vi electrica ortis, habita 1753 4to. Petrop.
CH. RABIQUEAU. Le Spectacle de la Nature du Feu élémentaire, ou cours d'Electricité expérimentale, où l'on trouve l'explication, la cause et le mechanisme du feu dans son origine, de là dans les corps, son action sur la bougie, sur le bois, etc. etc. 8vo. Paris.
E ROMAS. Neuer elektr. Versuch mit dem fliegenden Drachen am 14. Nov. 1753. Journ. d. Sav. Dec. 1753.
J. G. TESKEN's Abhandlung von dem Nutzen der Electricität in Abwendung des Ungewitters. Wöchentl. Königsb. Frag. und Anz. Nachr. No. 20 des J. 1753.
Lettre au R. P. J. sur une Expérience électrique. Journ. d. Sc. 1753. 241.
J. H. WINKLERI de avertendi Fulminis Artificio secundum Electricitatis doctrinam Commentatio. 4to. Lips.

1754. J. P. EBERHARD's Gedanken von den Ursachen der Gewitter und ihrer Aehnlichkeit mit der Elektricität. No. 31-33. Wöch. Hall. Anzeig. vom J. 1754.
E. M. FAIT. Observations concerning the Thunder and Electricity. Essays and Observations physical and chemical literary, read before a Society in Edinburgh. 189. Edinburgh.
J. LINING. Extract of a Letter from J. Lining of Charlestown, in South Carolina, to Charl. Pinckney in London, with his Answers to several Queries sent to him, concerning his Experiment in Electricity with a Kite. Phil. Trans. xlviii. 757.
B. FRANKLIN. New Experiments and Observations. London.

1755. T. MARINI de Electricitate coelesti, sive, ut alii vocant, naturali, Dissertatio. Commentar. de Bononiensi Scientiar. et Artium Instituto atque Academia, p. 205. Bonon. .
DE ROMAS. Mémoires, où après avoir donné un moyen aisé pour élever fort haut, et à peu de frais, un corps électrisable isolé, on rapporte des Observations frappantes, qui prouvent, que plus le corps isolé est élevé au-dessus de la terre, plus le feu de l'Electricité est abondant. Mém. de Math. et de Phys. prés. à l'Acad. à Paris, p. 393.
LE ROI. Mémoire sur l'Electricité résineuse, où l'on montre qu'elle est réellement distincte de l'Electricité vitrée, comme feu Mr. du Fay l'avoit avancé, et qu'elle nous fournit de nouvelles lumières sur les causes de l'Electricité naturelle et du Tonnerre. Mém. de Paris pour 1755, p. 264. Paris.
J. VERATTI. Dissert. de Electricitate coelesti. Comm. de Bononiensi Scientiar. et Artium Instituto atque Acad. p. 200. Bologna.
J. VERATTI. Nachricht von einem elektrischen Versuche mit dem Gewitter. Hamburger Magazin, xv. 602.

1756. E. M. FAIT. Beobachtungen vom Donner und der Elektricität. Aus d. Engl. in Edinb. neuen Versuche und Bemerk. aus der Arzneykunst und übrigen Gelehrs. i. 217. Altenburg.
DE ROMAS. Electrical Experiments made with a Paper Kite raised to a very considerable height in the Air. Gentleman's Magazine for Aug. 1756, 378.

1757. M. BUTSCHANY, Dissert. de Fulgure et Tonitru ex phaenomenis electris. 4to. Gotting.
TH. MARINI. Abhandlung von der himmlischen, oder wie Andere sie

nennen, natürlichen Elektricität. Allgem. Magaz. de Nat. Kunst und Wissenschaft. ix. 268. Leipzig.

1757. JNO. SMEATON. Effects of Lightning on a Steeple. Phil. Trans. l. 198.

DE ROMAS. Letter from M. de Romas to the Abbé Nollet, containing Experiments made with an Electrical Kite. Gentlem. Magaz. 109. March 1764.

J. VERATTI. Abhandlung von der himmlischen Elektricität. Aus. d. Lat. im Allg. Magaz. der Nat., Kunst u. Wissensch. ix. 261. Leipzig.

WILKE. Dissertatio de Electricitatibus contrariis. 4to. Rostock.

J. B. BECCARIA. Brief von der Elektricität an den Hrn. Abt Nollet gerichtet. Aus d. Franz. im Hamb. Mag. xviii. 378.

1758. DE ROMAS. Elektrischer Versuch mit einem sehr hoch in die Luft gestiegenen Papiernen Drachen. Aus dem Franz. im Brem. Magaz. iii. 114. Hannover.

B. FRANKLIN's Briefe von der Electricität. Aus d. Engl. von O. Wilke. Leipzig.

G. BECCARIA. Lettere dell' Elettricismo. 4to. Bologna.

1759. J. A. UNZER's Abhandlung vom Verhalten bei Gewittern, und von den Mitteln, die Gewitter, ehe sie noch reif werden, zu vernichten, oder wenigstens von einer Person und einem Hause abzuleiten. Medicin. Wochenschrift (der Arzt), i. 257. Hamburg.

J. C. WILKE. Von den Versuchen mit den eisernen Stangen, den Donnerschlag abzuwenden, und dem dabei beobachteten Merkwürdigsten. Abb. d. k. schwed. Akad. d. Wiss., deutsche Uebers. aus d. J 1759. xxi. 81.

HARTMANN. Von der Verwandtschaft und Aehulichkeit der elektrischen Kraft mit den erschrecklichen Lufterscheinungen. 8vo. Hannover.

1760. BARBERET. Abhandlung von der Aehnlichkeit, die sich zwischen den Erscheinungen bei dem Donner und der Elektricität findet, etc. etc. (Preisschrift). Aus d. Franz. im 1. St. des gemeinnütz. Natur- und Kunst- Magaz. 1. Berl.

J. LINING. Elektrische Versuche mit einem Papiernen Drachen. Aus d. Engl. im Hamburger Magaz. xxiv. 588.

1761. M. BUTSCHANY. Der Blitz entsteht nicht durch die Entzündung einiger brennbaren Theilchen, die in der Luft schweben, und es ist auch kein Feuer. Aus d. Lat. im 48. und 49. St. der Hannover. Beitr. z. Nutzen und Vergn. v. J. 1761.

1762. W. WATSON. Some Suggestions concerning the preventing the mischiefs which happen to Ships and their Masts by Lightning. Phil. Trans. lii. 629.

PET. VON MUSSCHENBROEK. Introductio ad Philosophiam naturalem. T. ii. Lugd. Bat. 4.

1763. J. F. HARTMANN. Gedanken über den Ursprung der Luftelektricität bei Gewittern. Im 55. und 56. St. des I. Jahrg. des Hannover. Magaz. v. J. 1763.

1764. T. BERGMANN. Tal om möjeligheten at förexomma åskans skadeliga werkningar. 4to. Stockholm.

Method of preserving Buildings from Lightning. Gentlem. Mag. June 1764, 284.

WILSON. Considerations to prevent Lightning from doing mischief to Great Works, High Buildings and Large Magazines. Phil. Trans. liv. 247.

WATSON. Some suggestions concerning Lightning-Storms. Phil. Trans. liv. 204.

MENASSIER. Elektr. Entladungen an Mastbäumen. Mem. d. l'Acad. R. d. Sc. an 1764, p. 408.

NOLLET. Mémoire sur les effets du Tonnerre comparés à ceux de l'Electr., avec quelques considérations sur les moyens de se garantir des premiers. Mém. de l'Acad. 1764.

1764. W. HEBERDEN. On the Effects of Lightning. Phil. Trans. liv. 198.
T. LAWRENCE. Effects of Lightning. Phil. Trans. liv. 235.
ED. DELAVAL. On the Effects of Lightning ou St. Bride's Church. Phil. Trans. liv. 227.

1765. Vorschlag, wie man Häuser vor dem Blitze bewahren könne. Aus d. Gentlem. Mag. 1764 übers. im Brem. Magaz. vii. 508.

1766. DALIBARD. Histoire abrégée de l'Electricité. 8vo. Paris.
PONCELET. La Nature dans la formation du Tonnerre. 8vo. Paris.

1767. B. FRANKLIN. Sur le Tonnerre et sur la Méthode que l'on employe communément aujourd'hui en Amérique pour garantir les hommes et les bâtimens de ses effets désastreux. Oeuvres i. p. 250. Paris.
LIND. Maison d'épreuve du petit Tonnerre. Oeuvres de Franklin, i. 302.
J. PRIESTLEY. The History and Present State of Electricity, with original Experiments. 4to. London.

1768. T. BERGMANN. Von der Möglichkeit, den schädlichen Wirkungen der Gewitter vorzubeugen. Aus d. Schwed. im Schwed. Magaz. i. 39, übers. v. C. Weber. Copenhagen.
J. C. LORIE. Ehre Gottes aus der Betrachtung des Himmels und der Erde. 8vo. Bd. v. Nürnberg.
P. DIVISCH. Längst verlangte Theorie von der meteorologischen Elektricität. 8vo. Frankfurt.
D. ROBERT. Von dem Abfluss der elektrischen Materie aus den Wolken in die Glocken. 233. Alton. gel. Merc. a. d. J. 1768.

1769. J. A. H. REIMARUS. Die Ursache des Einschlagens vom Blitze, nebst dessen natürlichen Abwendung von unseren Gebäuden, aus zuverlässigen Erfahrungen vor Augen gelegt. 8vo. Hamburg, 1768. Langensalza. 1769.
B. FRANKLIN. Experiments and Observations on Electricity, &c. &c. 4to. London.
D. J. G. KRÜNITZ. Von der natürlichen oder himmlichen Elektricität: Literatur. Verzeichn. d. vorn. Schriften, etc. etc. 8vo. 129. Leipzig.
NOLLET. Vergleichung der Wirkungen des Donners mit den Wirkungen der Elektricität, nebst einigen Betrachtungen über die Mittel sich vor dem Ersteren zu bewahren. Aus dem Franz. 8vo. Prag.
Royal Society. To secure St. Paul's from Lightning. Phil. Trans. lix. 160.

1770. BERTHOLON DE ST. LAZARE. Mémoires sur les Verges ou Barres métalliques destinées à garantir les édifices des Effets de la Foudre. Mém. de Par. 1770. 63. Electric. de Météores, i. 228.
J. PRIESTLEY. Additions on the History and present State of Electricity. 4to. London.
TODERINI. Filosofia Frankliana delle Punte preservatrici dal Fulmine. Modena.

1771. J. F. ACKERMANN. Programma, quo morbus et sectio fulmine nuper adusti enarratur. 4to. Kilæ.
D. J. P. EBERHARD. Vorschläge zur bequemeren und sichereren Anlegung der Pulver-Magazine. 8vo. Halle.
J. J. v. FELBIGER, Kunst- Thürme oder andere Gebäude vor den schädlichen Wirkungen des Blitzes durch Ableitungen zu bewahren, angebracht an dem Thurm der Sagan'schen Stifts- und Pfarrkirche. 8vo. Breslau.
SAUSSURE. Manifeste ou Exposition abrégée de l'utilité des Conducteurs Electriques. 8vo. Genève.

1772. J. F. Ackermann. Nachrichten von der sonderbaren Wirkung eines Wetterstrahles. 8vo. Kiel.
G. Beccaria. Della Elettricità terrestre atmosferica a cielo sereno. Osservazioni dedicate a sua Altezza Reale il Signor Principe di Piemonti. 4to. Torino.
P. Mako. Physikalische Abhandlungen von den Eigenschaften des Donners und den Mitteln wider das Einschlagen. Aus den Lat. von Retzer. 8vo. Wien.
J. Priestley. Geschichte und gegenwärtiger Zustand der Elektricität. Aus d. Engl. nach der 2. Ausg. von J. G. Krünitz. 4to. 9, 110, 206, 228, 254, etc. Berlin und Stralsund.
Mr. Henly. Effects of Lightning. Phil. Trans. lxii. 131.

1773. C. Steiglehner. Observationes phænomenorum electricor. in Hohen-Gebrachiu et Prifling. 4to. Ratisb.
Cavendish, Watson, and Franklin. A Report of the Committee appointed to consider a Method for securing the Powder Magazine at Purfleet from Lightning. Phil. Mag. lxiii. 42.
Wilson. Observations on Lightning and the Method of Securing Buildings from its effects. Phil. Trans. lxiii. 49.

1774. W. Henly. On Pointed and Blunt Conductors. Phil. Trans. lxiv.
J. N. Tetens. Ueber die beste Sicherung seiner Person bei einem Gewitter. 8vo. Bützow und Wism.
J. J. v. Felbiger. Die Kunst-Thürme und andere Gebäude vor den schädlichen Wirkungen des Blitzes durch Ableitungen zu bewahren. 8vo. Breslau.
Ph. P. Guden. Von der Sicherheit wider die Donnerstrahlen. 8vo. Götting. u. Gotha.
L. Ch. Lichtenberg. Verhaltungs-Regeln bei nahen Donnerwettern, nebst den Mitteln sich gegen die schädlichen Wirkungen des Blitzes in Sicherheit zu setzen. Zum Unterricht für Unkundige. 8vo. Gotha.

1775. Jos. Scudery. Fernglas der Arzeneiwissenschaft, nebst einigen anderen Abhandlungen, Schiffe und Häuser vor dem Blitze zu bewahren, ingleichen ganze Städte und Distrikte vor dem Erdbeben in Sicherheit zu setzen. Aus dem Ital. 8vo. Münster.
P. R. Arbuthnot. Abhandlung über die Preisfrage: Ob und was für Mittel es gebe, die Hochgewitter zu vertreiben etc. Abhandl. d. churfürstl. bayer. Akad. der Wiss. ix. 399.

1776. M. van Marum. Verhandeling over het Electrizeeren. 8vo. Groningen.
J. F. Gross. Elektrische Pausen. 8vo. Leipzig.
P. Cavallo. Extraordinary Electricity of the Atmosphere. Phil. Trans. lxvi. 407.
G. Beccaria. A Treatise upon Artificial Electricity. 8vo. London.

1777. F. X. Epp. Abhandlung von dem Magnetismus der natürlichen Electricität. 8vo. München, Kl.
W. Henly. Experiments and Observations in Electricity. Phil. Trans. lxvii. 85.

1778. L. Chr. Lichtenberg. Verhaltungs-Regeln bei nahen Donnerwettern, nebst den Mitteln, sich gegen die schädlichen Wirkungen des Blitzes in Sicherheit zu setzen. Zum Unterricht für Unkundige. 8vo. Gotha.
Nairne. Experiments in Electricity, being an Attempt to show the Advantage of Pointed Conductors. Phil. Trans. lxviii. 823.
J. A. H. Reimarus. Vom Blitze. 8vo. Hamburg.
J. Toaldo. Dei Conduttori per preservare gli Edifizii da Fulmini. 4to. Venez.

1778. BENJN. WILSON. On the Nature and Use of Conductors. Phil. Trans. lxviii.
Royal Society. Report of the Committee on Accident at Purfleet. Phil. Trans. lxviii.
B. WILSON. On the Termination of Conductors. Phil. Trans. lxviii. 909.
MR. BODDINGTON. Accident from Lightning at Purfleet. Phil. Trans. lxviii.
1779. B. TINAN. Mémoires sur les Conducteurs pour preserver les Edifices de la Foudre. 8vo. Strasbourg.
DR. INGENHOUSZ. New Experiments and Observations concerning Various Subjects. 8vo. London.
J. TOALDO. Mémoires sur les Conducteurs pour préserver les Edifices de la Foudre. Trad. de l'Italien, avec des notes et des additions par Barbier de Tinan. 8vo. Strasbourg.
B. FRANKLIN. Experiments on the Utility of long pointed Rods for securing Buildings from damage by Strokes of Lightning. Polit. Misc. and Phil. Pieces. 487. London.
1780. LORD MAHON. Principles of Electricity. 4to. Elmsly.
B. FRANKLIN. Sämmtliche Werke. Aus dem Englischen und Französischen übersetzt. Nebst des franz. Uebersetzers B. Dubourg Zusätzen und mit einigen Anmerk. versehen von G. T. Wenzel. Dresden.
1781. BERTHOLON. Mémoire ou Nouvelles Preuves de l'efficacité des Paratonnerres. Assemblée publique de la Soc. Royale de Montpellier. 60.
KERKHOF. Beschreibung einer Zurüstung, welche die anziehende Kraft der Erde gegen die Gewitterwolke und die Nützlichkeit der Blitzableitung sinnlich beweiset. 8vo. Berlin.
1783. J. J. HEMMER. Kurzer Begriff und Nutzen der Blitzableiter. 8vo. Mannheim.
J. J. HEMMER. Kurze und deutliche Anweisung, wie man durch einen an jedem Orte wohnenden Schmied, oder andere im Metall arbeitende Handwerker, eine sichere Wetterableitung mit sehr geringen Kosten an allerhand Gebäuden anlegen lassen kann. 8vo. Friedrichsstadt.
LUTZ. Unterricht vom Blitze und Wetterableitern. 8vo. Nüruberg.
1784. J. INGENHOUSZ. Vermischte Schriften physisch- medicinischen Inhaltes. Uebersetzt und herausgegeben von N. C. Molitor. Wien.
1785. J. PH. OSTERTAG. Archäologische Abhandlung über die Blitzableiter und die Kenntnisse der Alten von der Electricität. Neue philos. Abh. der baier. Akad. der Wiss. iv. 113.
J. HELFENZRIEDER. Verbesserung der Blitzableiter. 8vo. Eichstädt.
M. SIGAUD DE LA FOND. Précis historique et expérimental des Phénomènes Electriques. 8vo. Paris.
J. WEBER. Theorie der Electricität. Nebst Helfenzrieder's Vorschlag etc. 8vo. Salzburg.
M. LANDRIANA. Dell' utilità di Conduttori Elettrici. 4to. Milano.
1786. J. J. HEMMER. Anleitung Wetterableiter an allen Gattungen von Gebäuden auf die sicherste Art anzulegen. Mannheim und Frankfurt.
T. CAVALLO. A complete Treatise on Electricity. 8vo. London.
M. LANDRIANA. Abhandlung von Nutzen der Blitzableiter. Auf Befehl des Guberniums herausgegeben. Aus dem Italienischen von G. Müller. 8vo. Wien.
1787. BERTHOLON. De l'Electricité des Météores. 2 vol. Paris.
1788. A. PINAZZO. Diss. sopra alcuni buoni fisici Effetti che nascono da' Temporali. Mantova, Disser. 99.
1789. P. PLAC. HEINRICH. Abhandlung über die Wirkung des Geschützes auf Gewitterwolken. Gekrönte Preisschrift. Neue philos. Abh. der baier. Akad. der Wiss. v. I.

1790. Einige gegen die Gewitterableiter gemachte Einwürfe, beantwortet. 8vo. Frankfurt.
  BOKMANN. Beschreibung eines bequemen Apparates zur Beobachtung der Luftelektricität, nebst einigen Beob. und Versuchen. Gren Journ. i. 219. 385.
1791. BOECKMANN. Ueber die Blitzableiter. 8vo. Karlsruhe.
  BUSSE. Beruhigung über die neuen Blitzableiter. 8vo. Leipzig.
  C. G. VON ZENGEN. Ueber das Läuten bei Gewittern, besonders in Hinsicht der deshalb zu treffenden Polizeyverfügungen. 8vo. Giessen.
  DE LUC. Ueber das Elektrische Fluidum. Gren Journal de Phys. iii. 91.
  H. MEURER. Abhandlung von dem Blitze und den Verwahrungs-Mitteln gegen denselben. 4to. Trier.
1792. BERTHOLON. Von der Elektricität der Lufterscheinungen. Deutsch von —— Liegnitz. 8vo.
  F. A. WEBER. Abhandlung vom Gewitter und Gewitterableiter. Zürich.
  J. W. WALLOT und CASSINI. Beobachtungen über die Oscillationsbewegung der Magnetnadel unmittelbar nach dem Vorüberziehen eines Gewitters. Gren Journ. v. 83.
1793. A. VOLTA. Meteorologische Briefe. Aus d. Italien. 8vo. Leipzig.
1794. J. A. H. REIMARUS. Neuere Bemerkungen vom Blitze, dessen Bahn, Wirkung, sicheren und bequemen Ableitung. Aus zuverlässigen Wahrnehmungen von Wetterschlägen dargelegt. 8vo. Hamburg.
  J. A. H. REIMARUS. Ausführliche Vorschriften zur Blitz-Ableitung an allerlei Gebäuden. Aufs Neue geprüft etc. 8vo. Hamburg.
1795. CHAPPE. Ueber die Eigenschaft der Spitzen, elektr. Materie aus bedeutenden Entfernungen aufzunehmen. Gren n. Journ. d. Phys. i. 115.
1796. J. F. GROSS. Grundsätze der Blitzableitungskunst. Nach dem Tode des Verf. herausgegeben von J. F Wiedemann. 8vo. Leipzig.
1797. T. CAVALLO. Vollst. Abhandl. der Lehre v. d. Elektr. Aus d. Engl. 4. Ausgabe. 8vo. Leipzig.
  K. G. KÜHN. Die neuesten Entdeckungen in der Elektr. 2 Theile. ii. 1—173. Leipzig.
1798. FR. K. ACHARD. Kurze Anleitung, ländliche Gebäude vor Gewitterschäden sicher zu stellen. 8vo. Berlin.
1799. A. VOLTA. Meteorologische Beobachtungen, besonders über die atmosphärische Elektricität. Aus d. Italienischen mit Anmerkungen des Herausgebers. (Herausgeg. von Lichtenberg, übers. von Schäffer.) 8vo. Leipzig.
  VAN MARUM. Versuche für Blitzableiter. Gilbert's Ann. i. 263.
1800. V. HAUCH. Von der Luftelektricität, besonders mit Anwendung auf Gewitterableiter. Kopenhagen.
  H. HALDANE. Versuche, den Grund zu entdecken, weshalb der Blitz in Gebäude einschlug, die mit Gewitter-Ableitern versehen waren. Gilbert's Ann. v. 115. Nicholson's Journal of Natur. Philos. i. 433.
  L. A. V. ARNIM. Einige Elektrische Bemerkungen. Gilbert's Ann. vi. 110.
1801. WOLFF. Versuche über Blitzableiter. Gilbert's Annalen, viii. 69.
1802. J. A. EITELWEIN. Kurze Anleitung, auf welche Art Blitzableiter an den Gebäuden anzulegen sind. 8vo. Berlin.
  GEORG CHRISTOPH LICHTENBERG. Ueber Gewitterfurcht und Blitzableitung. 8vo. Göttingen.
1803. G. CH. LICHTENBERG. Neueste Geschichte der Blitzableiter. Aus d. Jahre 1779. Math. und Phys. Schriften etc. i. 210.
  G. CH. LICHTENBERG. Vorschlag den Donner auf Noten zu setzen. Math. und Phys. Schriften etc. i. 478.

1803. G. Ch. Lichtenberg. Versuche zur Bestimmung der zweckmässigsten Form der Gewitterstangen. Math und Physik. Schriften, iii. 3.
1804. Michaelis und Lichtenberg's Briefwechsel über die Absicht oder Folgen der Spitzen auf Salomon's Tempel. Math. und Physik. Schriften, iii. 251.
Bodde. Grundzüge zu der Theorie der Blitzableiter. 8vo. Münster.
J. F. Lutz. Lehrbuch der theoretischen und practischen Blitzablitungslehre. Neu bearbeitet von J. K. Gütle. 2 Thle. 8vo. Nürnberg.
Saxtorph's Elektricitätslehre. 2 Theile. ii. 1—101. Kopenhagen.
1805. J. K. Gütle. Allgemeine Sicherheitsregeln für Jedermann bei Gewittern. Merseburg.
W. A. Lampadius. Versuche und Beobachtungen über Elektricität und Wärme der Atmosphäre. 8vo. Leipzig.
W. A. Lampadius. Ein Schneegewitter, und ein Vorschlag zur Vervollkommung der Blitzableiter. Gilbert's Ann. der Physik, xxix. 58.
1809. J. J. Hemmer. Der Rathgeber, wie man sich vor Gewittern in unbewaffneten Gebäuden verwahren soll. 8vo. Mannheim.
Bodde. Grundzüge zur Theorie der Blitzableiter. 8vo. Münster.
1810. J. Ph. Ostertag. Antiquarische Abhandl. über Gewitterelektricität. Auswahl aus den kl. Schriften des ..... Sammlung ii. 455. Salzbach.
J. A. H. Reimarus. Ueber die Sicherung durch Blitzableiter. Gilbert's Ann. xxxvi. 113.
1811. L. von Unterberger. Nützliche Begriffe von den Wirkungen der Elektricität und der Gewittermaterie, nebst einer practischen Belehrung 8vo. Wien.
M. v. Imhof. Ueber das Schiessen gegen heranziehende Donner- und Hagelgewitter. 4to. München.
B. Cook. On the Prevention of Damage by Lightning. Nicholson's Journal of Philosophy and Chemistry. Aug. 1811, and Feb. 1812.
1812. J. K. Gütle. Neue Erfahrungen über die beste Art Blitzableiter anzulegen. 8vo. Nürnberg.
1814. G. J. Singer. Elements of Electricity and Electro-Chemistry. 8vo. London.
1815. ———— Ueber Blitzableiter aus Messingdraht. Anzeiger für Kunst- und Gewerbefleiss in Bayern. No. 7. 81. München.
Benzenberg. Nachrichten über das Gewitter vom 11. Jan. 1815. Gilbert's Ann. l. 341.
Bodde. Ueber Blitzableiter. Gilbert's Ann. li. 80.
1816. M. v. Imhof. Theoretisch practische Anweisung zur Anlegung zweckmässiger Blitzableiter. 8vo. München.
Ueber Blitzableiter. Anzeiger für Kunst- und Gewerbefleiss in Bayern. No. 26. 418. München.
1818. C. A. W. Wenzel. Ueber Blitzableiter. Aus d. Französ. (?) Wesel.
1820. ———— Nothwendigkeit der Blitzableiter. Kunst- und Gewerbeblatt f. d. Königreich Bayern. Jahrg. 1820. No. 21. 166.
F. Trechsel. Bemerkungen über Blitzableiter und Blitzschläge, veranlasst durch einige Ereignisse im Sommer 1819. Gilbert's Ann. lxiv. 227.
La Postolle. Traité des Parafoudres et des Paragrêles. 8vo. Amiens.
1821. La Postolle. Ueber Blitz- und Hagelableiter aus Strohseilen. Aus d. Französ. Mit einer Abbildung. 8vo. Weimar.
Gay-Lussac's Bericht über La Postolle's Blitzableiter aus Stroh. Gilb. Ann. lxviii. 210.
Müller und Hofmann. Einige prüfende Versuche hierüber. Gilbert's Annalen, lxviii. 218.
Lindner. Blitzableiter von Strohseilen. Magazin der neuesten Erfind-

ungen, Entdeckungen und Verbesserungen von Poppe, Kühn und Baumgärtner. Neue Folge, ii. 18.

1821. VINCENT. Blitzableiter von Stroh. Journ. d. Connaiss. Usuell. et Pratiques, et Recueil des Notions etc., par Gillet de Grandmont et C. de Lasteyrie et d'autres. 8vo. xix. 281. Paris.

1822. DAVY. Neue tragbare Blitzableiter. Polyt. Journ. ix. 133.

WEBER. Die Sicherung unserer Gebäude durch Blitzstrahlableiter, theor. und pract. beleuchtet und bewährt, sammt einer Beurtheilung der Ableiter aus Stroh. Landshut.

1823. HARRIS. Observations on the Effect of Lightning on Floating Bodies; with an Account of a New Method of applying Fixed and Continuous Conductors of Electricity to the Masts of Ships. 8vo. London.

HARRIS. Ueber den Nutzen der Blitzableiter in der Oeconomie. Polyt. Journ. x. 372.

GAY-LUSSAC. Instruction sur les Paratonnerres. Ann. de Ch. et de Phys. xxvi. 258. Poggendorff's Ann. i. 403. (S. Seite 203.)

J. C. V. YELIN. Ueber den am 30. April 1822 erfolgten merkwürdigen Blitzschlag auf den Kirchthurm zu Rossstall im Rezatkreise, Bayern. 8vo. München.

1824. J. C. V. YELIN. Dasselbe. Auch unter dem Titel: Ueber die Blitzableiter aus Messingdrahtstricken etc. 8vo. 2. vermehrte Auflage. München.

ZIEGLER. Blitzableiter von Platina. Allgem. Handluugszeit. v. Leuchs. 175. Ann. de l'Indust. nation. et étrang. etc. xviii. 320.

1825. FISCHER. Ueber die Nachtheile magnetischer eiserner Ableitungsröhren. Kastner's Archiv f. d. gesammte Naturlehre, iii. 421.

PFAFF. Ueber Blitz und Blitzableiter. Gehler's physikalisches Wörterbuch, neu bearbeitet von Brandes, Gmelin, Horner, Muncke und Pfaff Bd. i. Abth. 2. 981—1003. Leipzig.

1827. HEUL. Anleitung zur Errichtung und Untersuchung der Blitzableiter für Bauverständige, Bau- und Feuerbeschauer und Gebäude-Inhaber. Stuttgart.

1828. MURRAY. Treatise on Atmospheric Electricity, including Observations on Lightning-Rods. 8vo. London.

R. HARE. Ueber die Ursachen, warum Wetterableiter in einigen Fällen nicht schützen, und die Mittel, dieselben vollkommen schützend zu machen, nebst einer Widerlegung der herrschenden Idee, dass Metalle die Elektricität vorzüglich anziehen. Aus Gill's Technological Repository. Nov. 1827, im Polyt. Journ. xxvii. 208.

1830. D. BREITINGER. Instruction über Blitzableiter im Canton Zürich. 4to Zürich.

BÖCKMANN. Ueber Blitzableiter. Eine Abhandl. auf höchsten Befehl bearbeitet. Neue Aufl. von Wucherer. Karlsruhe.

POPPE. Gewitterbuchlein zum Schutz und zur Sicherstellung gegen die Gefahren der Gewitter, besonders auch über die Kunst, Blitzableiter auf die beste Art anzulegen. Stuttgart.

PREINSCH. Ueber Blitzableiter, deren Nutzbarkeit und Anlegung. 8vo. Leipzig.

HARRIS. On the Utility of fixing Lightning Conductors on Ships. 8vo. Plymouth.

1831. MURRAY. Treatise on Atmospheric Electricity &c., traduit par Riffault. Paris.

BLESSON. Verbesserung an Blitzableitern. Verhandl. des Vereins zur Beförderung des Gewerbefleisses in Preussen. Jahrg. 1831. 260.

W. S. HARRIS. Ueber Blitzableiter an Schiffen. Aus Register of Arts. Oct. 1831. 211, in Polyt. Journ. xlii. 415.

1832. L. F. KÄMTZ. Von den elektrischen Erscheinungen der Atmosphäre. Lehrbuch der Meteorologie. Bd. ii. Abschnitt vii. 389. Halle.
1833. A. DE TAVERNIER. Blitzableiter, genannt Antijupiter, oder Tavernier's Gewitterableitende Säule. 8vo. Leipzig.
G. MAYR's Abhandlung über Elektricität und sichernde Blitzableiter für jedes Gebäude, für Reise- und Frachtwagen, Schiffe und Bäume. 8vo. München.
1834. P. BIGOT. Anweisung zur Anlegung, Construction und Voranschlagung der Blitzableiter für angehende Baubeamte, Bauhandwerker, insbesondere Metallarbeiter, und zunächst Hauseigenthümer und Oekonomen. Glogau.
J. HANCOCK. On the Cause of Heat Lightning. Phil. Mag. iv. (s. 3), 340
1835. PLIENINGER. Ueber die Blitzableiter. 8vo. Stuttgart und Tübingen.
Legirung für Blitzableiterspitzen. Journ. des Conn. xxii. 129. Polyt. Journal, lviii. 479.
1837. MARTYN ROBERTS. On Lightning Conductors, particularly as applied to Vessels. Read before the Electrical Society of London, June 24, 1837. Annals of Electricity, i. 468. 8vo.
K. W. DEMPF. Ueber Blitzableiter. Förster's Bauzeitung. Jabrg. 1837.
1838. ARAGO. Sur le Tonnere. Annuaire du Bureau des Longit. pp. 249, 255, 257, 451. Paris.
P. RIESS. Zusammenstellung der neueren Fortschritte über Atmosphärische Elektricität. Repertorium der Physik, ii. 87.
W. SNOW HARRIS. On the Protection of Ships from Lightning. Annals of Electricity, ii. 81.
MARTYN ROBERTS. Reply to W. Snow Harris's Paper on Lightning Conductors. Ib. ii. 241.
Mr. STURGEON. On the Principle and Action of Lightning Conductors. A Paper read before the London Electrical Society, March 7. Ib. ii. 383.
J. MURRAY. Lightning Rods. Ib. iii. 64.
1839. WM. STURGEON. On Marine Lightning Conductors. Addressed to the British Association, Birmingham Meeting, Sept. 8. Ib. iv. 161.
WM. STURGEON. Supplementary Note on Marine Lightning Conductors. Ib. iv. 235.
W. SNOW HARRIS. On Lightning Conductors, &c., being an Investigation of Mr. Sturgeon's Experimental and Theoretical Researches in Electricity, published by him in the Annals of Electricity, &c. Ib. 310.
W. S. HARRIS. On Lightning Conductors. Phil. Mag. xv. (s. 3), 461.
M. BREONOT. On the Thunder Clap which struck the Dome of the Hôtel des Invalides, June 8. Compt. Rend. June 17.
BÖCKMANN. Ueber Blitzableiter. 3. Auflage von G. F. WUCHERER. 8vo. Carlsruhe.
1840. Report of the Committee appointed by the Admiralty to examine the Plans of Lightning Conductors. Sturgeon's Ann. v. 1.
LIEUT. GREEN. On Lightning Conductors. Ib. iv. 330.
W. S. HARRIS. On Lightning Conductors, and the Effects of Lightning on certain Ships in H. M. Navy. Phil. Mag. xvi. (s 3.), 116, 401.
STURGEON. On the Subject of Marine Lightning Conductors. Annals of Electricity, iv. 496.
W. S. HARRIS. On Lightning Conductors. Ib. v. 208.
W. STURGEON. On Marine Lightning Conductors. Ib. v. 53, 220.
1841. M. FARADAY. On some supposed Forms of Lightning. Phil. Mag., June 22.
W. S. HARRIS. On Lightning Conductors, and on Experiments relating to the Defence of Shipping from Lightning. Phil. Mag. xviii. (3 s.), 172.

1841. J. MURRAY. On Lightning Conductors. Annals of Electricity, vii. 82.
J. ARROWSMITH. On the Use of Black Paint, in averting the Effects of Lightning on Ships. The Transactions and the Proceedings of the London Electrical Society from 1737 to 1840, p. 103. 4to. London.
W. STURGEON. A Paper on the Principle and Action of Lightning Conductors. Ib. p. 142.
W. L. WHARTON. The Effect of a Lightning Stroke. Ib. p. 102.
LUOTSKY. Some Remarks on Lightning on the High Seas. (Abstract.) Ib. 174.
W. STURGEON. Electric Storms. Annals of Electricity, vii. 400.

1842. C. V. WALKER. On Lightning Conductors. Phil. Mag. xxi. (3 s.), 63, 310.
W. S. HARRIS. Observations on a Paper by C. V. Walker, entitled, On the Action of Lightning Conductors. Phil. Mag. xxi. (3 s.), 313.
W. STURGEON. Description of a Thunderstorm as observed at Woolwich; with some Observations relative to the cause of the deflection of Electric Clouds by High Lands; and an Account of the Phenomena exhibited by means of a Kite elevated during the Storm. Annals of Electricity, ix. 167.
CHANTRELL. Ueber Blitzableiter. Polyt. Journ. lxxxvi. 179.
P. RIESS. Ueber Atmosphärische Elektricität und Schutzmittel gegen Elektrische Meteore. Repertorium d. Physik, vi. 277.
K. W. DEMPP. Vollständiger Unterricht in der Technik der Blitzableitersetzung nach 66 Modellen. Kl. 8. München.

1843. C. V. WALKER. Memoir on Lightning Flashes. Phil. Mag. xxii. (3 s.), 490.
W. B. O'SHAUGHNESSY. On the use of Lightning Conductors in India. Phil. Mag. xxiii. (3 s.), 177.
W. S. HARRIS. On the Nature of Thunderstorms, and the Means of Protecting Buildings and Shipping against Lightning. 8vo. London.

1844. M. A. FARGEAUD. On Lightning Conductors at Strasbourg Cathedral. Builder, ii. 39.
Edinburgh Review. On the best Method of Protecting Buildings from Lightning. (Abstract.) Builder, ii. 550.
MR. WALKER. On the Lightning Conductor on Royal Exchange. Builder, ii. 573.

1845. MR. WHITE. On Protection of Buildings from Lightning. Builder iii. 413.
CORNAY. Sur quelques Effets de l'Ouragan du 19 août observés dans les Environs de Paris. Compt. Rend. xxi. 534.
MARIANINI. Du ré-électromètre comme moyen de découvrir la direction de la Foudre. Ann. de Chim. et de Phys. xiii. 245.
HENRY. Method of Protecting from Lightning Buildings covered with Metallic Roofs. Proceed. of the Americ. Philos. Soc. iv. 179.

1846. F. REICH. Electrische Versuche I. Abh. b. Begründ. der Königl. Sächs. Gesellsch. der Wiss. p. 197.
STRICKER. Ueber Anwendung des Galvanismus zur Prüfung der Blitzableiter. Pogg. Ann. lxix. 554. Polyt. Journ. ciii. 265.
HENRY. Ueber ein einfaches Verfahren, Gebäude mit metallischer Bedeckung vor dem Blitz zu schützen. Polyt. Journ. ci. 43.

1847. OLMSTED. A New Effect of the Magnetic Telegraph. Mech. Mag. xlvii. 262.
A. KUNZEK. Atmosphärische Elektricität. Leicht fassliche Darstellung der Meteorologie. Wien. Gr. 8. 174.
Mr. HIGHTON. A Paper on Lightaing Conductors, read before Society of Arts. Builder, v. 18.

1847. W. SMITH. On the Protection of Buildings from Lightning. Builder, v. 190.
M. BONJEAN. On the Presence of Sulphur in Substances struck by Lightning. Phil. Mag. xxx. (3 s.), 222.
HARRIS. On some Recent and Remarkable Examples of the Protection afforded by Metallic Conductors against heavy Strokes of Lightning. Proc. British Assoc. 1848.

1848. ISHAM BAGGS. On the Proximate Cause of Lightning. Proceedings of Royal Society, v. 731.
H. POSELGER. Berichte über atmosphärische Elektricität aus d. J. 1846. Fortschritte d. Physik, herausgegeben von d. Phys. Gesellsch. zu Berlin (oder Berl. Ber.) f. 1848. 8vo. p. 363. Berlin.
C. BRUNNER. Elektrische Lichterscheinungen ohne Donner. Fror. Not. ix. x. p. 152.
LADAME. Sur les Phénomènes électriques de l'Air. Bibl. Univ. ix. 286.
W. EISENLOHR. Anleitung zur Ausführung und Visitation der Blitzableiter. 8. Karlsruhe.

1849. T. H. DIXON. On Rain, the Cause of Lightning. Phil. Mag. xxxv. (3 s.), 392.
R. BIRT. On the Production of Lightning by Rain. Phil. Mag. xxxv. (3 s.), 161.
PORRO. Bleiröhren für Blitzableiter. Polyt. Journ. cxv. 397.
E. HIGHTON. Action perturbatrice de l'Electricité atmosphérique. Compt. Rend. xxix. 126.
MORLET. Resultats de Recherches nouvelles sur l'Arc lumineux qui accompagne souvent les Aurores Boréales. Compt. Rend. xxviii. 744, 789.
DE LA RIVE. Sur les Aurores Boréales. Arch. d. Sc. Phys. et Nat. xii. 222.

1850. R. PHILIPPS. On the Connection of the Electricity of Condensation with Lightning and the Aurora. Phil. Mag. xxxvi. (3 s.), 103.
W. R. BIRT. On the Connection of Atmospheric Electricity with the Condensation of Vapour. Phil. Mag. xxxvi. (3 s.), 161
W. R. BIRT. Ueber die Veränderungen der Messingdrahtseile bei Blitzableitern. Bayer. Kunst und Gewerbebl. 148.
PELTIER. Sur l'Electricité atmosphérique. Bull. d. Brux. xvii. 1. p. 5.
PORRO. Substitution d'un Tube de Plomb à la Corde métallique communément employé comme Conducteur pour les Paratonnerres. Compt. Rend. xxx. 86.
PORRO. Deuxième Note sur les Paratonnerres. Institut. 149.
E. HIGHTON. Action de l'Electricité atmosphérique sur les Télégraphes électriques. Institut. 80.
CH. FR. SCHÖNBEIN. Ueber den Ursprung der Wolkenelektricität und der Gewitter. In der Denkschrift: Ueber den Einfluss des Sonnenlichtes auf die Chemische Thätigkeit des Sauerstoffes und den Ursprung der Wolkenelektricität und des Gewitters. 11. Basel.
R. PHILIPPS. On the Theory of Thunder-Storms. Phil. Mag. xxxvii. (3 s.), 510.
J. P. JOULE. On a Remarkable Appearance of Lightning. Phil. Mag. xxxvii. (3 s.), 127.
ED. LOWE. Observations on 287 Thunder-Storms. Proceedings of Royal Society, v. 957.
PETER CLARE. On some Thunder-Storms and Extraordinary Electrical Phenomena. Phil. Mag. xxxvii. (3 s.), 329.
Prof. WM. THOMSON. On some Remarkable Effects of Lightning. Ib. (3 s.), 53.

1851. M. QUETELET. On Atmospheric Electricity. Ib. i. (s. 4), 329.

1851. E. LOOMIS. On the proper Height of Lightning Rods. Silliman's Journ. (2), x. 320.

W. STURGEON. On Lightning and Lightning Conductors. Mem. of the Manch. Soc. (2), ix. 56.

CASASECA. Cas de Foudre observé à la Havane. Compt. Rend. xxxiii. 200.

J. LAMONT. Messung der Atmosphärischen Elektricität. Abhandl. dre Math. Physik. Cl. d. k. b. Akademie d. Wiss. vi. 2. p. 437.

J. LAMONT. Beobachtungen der Luft-Elektricität an der Münchener Sternwarte vom 1. Mai 1850 bis Ende October 1851. Pogg. Ann. lxxxv. 494.

ARNOLD. Blitzableiter zum Schutz der Wärterbuden. Polyt. Centralblatt. 650.

1852. H. POSELGER und G. KARSTEN. Berichte über 'Atmosphärische Elektricität.' Berl. Ber. 1848. 275. Berlin.

A. D'ABBADIE. Sur les Orages d'Ethiopie. Compt. Rend. xxxiv. 894.

W. HAIDINGER. Niedrigste Höhe der Gewitterwolken. Wiener Sitzungsberichte, iv. 338.

K. FRITSCH. Die tägliche Periode der Gewitter und ihre Ursache. Wiener Sitzungsberichte, ix. 800.

M. QUETELET. On Atmospheric Electricity. Phil. Mag. iv. (4 series), 249.

R. PHILLIPS. On the Electrical Condition of the Atmosphere. Phil. Mag. iv. (4 s.), 126.

1853. J. SPRATT's. Faugstange für Blitzableiter. Polytechn. Centralbl. 1142.

BEETZ. Berichte über Atmosphärische Elektricität. Berl. Ber. f. 1849. 258. Berlin.

ARMITAGE. Lightning Rod. Mech. Mag. lix. 204.

E. B. BRIGHT. Lightning Conductors. Mech. Mag. lix. 246.

C. BECK. Einige Worte über Blitzableiter. Zeitschrift f. die gesammten Natur-Wissenschaften, ii. 220.

P. TH. RIESS. Entladungs-Erscheinungen der Atmosphärischen Elektricität. Die Lehre von der Reibungs-Elektricität. Gr. 8. Bd. ii, Kap. 3. 528. Berlin.

1854. F. ARAGO. Le Tonnerre. Oeuvres de F. Arago. Notes scientifiques I. Cosmos. V. 30, 700. Edinb. Journ. (2), iii. 150.

F. ARAGO. Ueber das Gewitter. Arago's sämmtliche Werke. Mit einer Einleitung von Alexander v. Humboldt. 8vo. Leipzig.

F. COHN. Ueber die Einwirkungen des Blitzes auf die Bäume. Jahresberichte der schlesisch. Gesellsch. 1853, p. 1.

T. DU MONCEL. Théorie des Eclairs. Mém. de la Soc. de Cherbourg. ii. 49.

LECLERQ. Sur la Cause qui produit le Bruit prolongé du Tonnerre. Compt. Rend. xxxix. 694.

POUILLET. Supplément à l'Instruction sur les Paratonnerres.

C. DUPIN. Observations au sujet du Rapport sur l'Etablissement de Paratonnerres à bords des Vaisseaux. Compt. Rend. xxxix. 1150.

NASMYTH. FARADAY. On Lightning Conductors. Athenæum (1854), 1182.

J. L. GATCHELL. Lightning Rod. Mech. Mag. lxi. 174.

R. B. FORBES. Lightning Conductors for Ships. Mech. Mag. lxi. 178.

WITTCKE. Ueber das Gewitter. Vorgelesen am 1. April 1844 in der Sitzung der Erfurter Akademie gemeinnütziger Wissenschaften. S. Cassel's Wissenschaftliche Berichte, ii.—iii. 68. Erfurt.

1855. J. LAMONT. Berichte über Atmosphärische Elektricität. Berl. Ber. 1850—1851. 879. Berlin.

W. S. HARRIS. Protection of the New Palace of Westminster from Lightning. Mech. Mag. lxii. 302.

1855. F. Arago. Meteorological Essays. Translated by Colonel Sabine; with an Introduction by Baron von Humboldt. 8vo. London.
Colonel Sabine. On Thunderstorms. Proc. Royal Society, vol. vii. 347.

1856. Dellmann. Berichte über Atmosphärische Elektricität, 612. Ber. Ber. 1853. Berlin.
Becquerel. Recherches sur l'Electricité de l'Air et de la Terre, et sur les effets chimiques produits en vertu d'actions lentes avec ou sans le concours des forces électriques. Compt. Rend. xliii. 1101.
S. Mästermann. Observations on Thunder and Lightning. Smithsonian Report for 1855, 265.
Baillard. Sur les Eclairs sans Tonnerre et les Tonnerres sans Eclairs. Compt. Rend. xliii. 816.
Lenz. Sur combien de pieds carrés de la surface de la toiture doit-on, en construisant un Paratonnerre, établir un Conducteur à terre?—Bullet de la Classe phisico-mathématique de l'Acad. Impériale de St.-Pétersbourg, xv. 63.
J. Müller. Atmosphärische Elektricität. Lehrbuch der kosmischen Physik. 8vo. Braunschweig.
C. S. M. Pouillet. Eléments de Physique expérimentale et de Météorologie. 7th edition, 2 vols. 8vo. Paris.
Guiot. Sur la substitution du Cuivre au Fer. Compt. Rend. xliii. 1205.

1857. Babinet. Ib. Compt. Rend. xliv. 636.
Count du Moncel. Note on Thunder and Lightning. Compt. Rend. 49.

1858. M. Ronneau. Paratonnerres. De leur emploi pour mettre les cultures à l'abri de la grêle. Compt. Rend. xlvi. 589, 743.
M. Pouillet. Rapport fait à l'Académie sur la question de Paratonnerres. Compt. Rend. xlvii. 287.
M. Pimenta. Sur un Nouveau Système de Paratonnerre. Compt. Rend. 157.

1859. G. A. Rowell. An Essay on the Cause of Rain and its allied Phenomena. 8vo. Oxford.
C. Tomlinson. The Thunderstorm. 8vo. London.

1861. Duret. Lettre sur un cas d'inefficacité des Paratonnerres. Compt. Rend. liii. 23.
Guiot. Sur les Indications à remplir dans l'installation des Paratonnerres. Compt. Rend. liii. 290.

1862. Pouillet. Rapport sur le Coup de Foudre qui a frappé le Magasin à Poudre, Place de Bethune, le 16 Juin 1862. Compt. Rend. lv. 267.
C. Tomlinson. On Lightning Figures. British Association Report, 1862.
Sacré. Sur la Construction des Paratonnerres. Compt. Rend. lv. 444.
Callaud. Lettre sur Certaines Dispositions qu'il donne aux Paratonnerres au but d'en augmenter l'Efficacité. Compt. Rend. lv. 697.
Perrot. Note sur les Résultats d'Expérience entreprises dans le but d'accroitre l'efficacité de ces Appareils. Compt. Rend. liv. 852.
Perrot. Note sur les Moyens d'augmenter l'Efficacité des Paratonnerres. Compt. Rend. lv. 361, 465.
Perrot. Sur les Paratonnerres armés d'une couronne de Pointes aiguës. Compt. Rend. lv. 642.

1863. Perrot. Nouvelles Expériences tendant à prouver que lorsqu'un Paratonnerre ordinaire est foudroyé, son Conducteur devient foudroyant pour les corps voisins. Compt. Rend. lvi. 397.
Perrot. Note sur les Rapports des Distances auxquelles s'etendent les actions neutralisantes de la Pointe du Paratonnerre ordinaire et d'une Pointe très-effilée. Compt. Rend. lviii. 115.

1865. Melsens. Sur les Paratonnerres à Conducteurs multiples. Compt. Rend. lxi. 84.

1866. CARL KUHN. Handbuch der angewandten Elektricitätslehre. Part I. Ueber Blitzableiter. 8vo. Leipzig.
M. BOUDIN. On Deaths by Lightning. The Year Book of Facts. 8vo. London.
1867. DR. OTTO BUCHNER. Die Konstruktion und Anlegung der Blitzableiter 8vo. Weimar.
L. FIGUIER. Les Merveilles de la Science. 4to. Paris.
M. BALTARD. Consulte l'Académie des Sciences de France relativement aux Dispositions adoptées pour les Paratonnerres de l'Eglise St. Augustin. Compt. Rend. lxv. 453.
M. PELTIER. On Lightning Conductors. Proc. Belgian Academy of Sciences, 1867.
POUILLET. Projet d'Instructions sur les Paratonnerres, préparé pour répondre à une demande de M. le Ministre de la Guerre. Compt. Rend. lxiv 80, 182.
1868. LIEUT JOHN HERSCHEL, R. E. On the Lightning Spectrum. Philos. Mag. xxxvii. (4 s.), 142.
M. LEFUEL. Rapport concernant les Paratonnerres des Tuileries et du Louvre. Compt. Rend. lxvi. 415.
1869. M. MELSENS. Notice sur le Coup de Foudre de la Gare d'Anvers du 10 juillet 1865. Mémoires couronnés, Acad. Royale de Belgique, xxvi. 1875.
W. DE FOUVIELLE. Éclairs et Tonnerres. 8vo. Paris.
M. POUILLET. Instruction sur les Paratonnerres du Louvre et des Tuileries, rédigée au nom d'une Commission par feu M. Pouillet, lue et approuvée par l'Académie des Sciences. Compt Rend. lxvii. 148.
M. DE PARVILLE. Note sur un Procédé de Contrôle de la Conductibilité des Paratonnerres. Ib. 306.
M. VAILLANT. Un travail relatif aux mesures qui ont été prises pour les Magasins à Poudre de France et d'Algérie en ce qui concertne les Paratonnerres. Ib. lxviii. 700.
M. BECQUEREL. On the Return Stroke of Lightning. Mechanics' Magazine, London.
M. ANICH. On the Influence of Local Agents on the Production of Thunderstorms. Philos. Mag. xxxviii. (4.s.) 436.
1871. HERMANN J. KLEIN. Das Gewitter, und die Mittel sich vor den Verheerungen des Blitzes zu schützen. 8vo. Gratz.
C. A. JOHNS. On Thunderstorms. 'Nature,' iv. 367.
1872. DR. WILHELM STRICKER. Der Blitz und seine Wirkungen. 8vo. Berlin.
HENRY WILDE. On the influence of Gas and Water-Pipes in determining the Direction of a Discharge of Lightning. Philos. Mag. vol. xliii. (4.s.) 115.
J. P. JOULE. On the Spectrum of Lightning. 'Nature,' vi. 186.
M. W. DE FOXVIELLE. The Efficiency of Lightning Conductors. Compt. Rend. No. 15. Oct. 17, 1872.
SECCHI. Phenomena produced by Lightning. Telegraph. Journal. vol. i. 25.
PROF. C. V. ZENGER. On Symmetric Conductors and the Construction of Lightning Conductors. Brit. Assoc. Report, 1873, 41.
M. D. COLLADON. Effects of Lightning on Trees. Compt. Rend. 19. Nov. 1872.
W. H. PREECE. On Lightning and Lightning Conductors. Jour. Soc. Tel. Eng. i. 336.
JAS. GRAVES. On Lightning Conductors. Ib. p. 413.
1873. W. DE FOXVIELLE. Note sur un Projet de Paratonnerres à Condensateur. Compt. Rend. lxxvi. 384.

1873. W. DE FONVIELLE. Sur Différents Mouvements électriques observés sur le Paratonnerre interrompu de l'Observatoire de Greenwich. Ib. 1282.

W. DE FONVIELLE. Sur les Causes multiples qui provoquent la chute de la Foudre. Ib. 1394.

W. DE FONVIELLE. On the Advantages of keeping Records of Physical Phenomena connected with Thunderstorms. Brit. Assoc. Report, 55. 1873.

JNO. M. MOTT. Lightning and Lightning Rods. Journal of the Franklin Institute. 8vo.

E. GRENET. Construction de Paratonnerres. 8vo. Paris.

PROFESSOR REYNOLDS. On the Electrical Properties of Clouds and the Phenomena of Thunderstorms. Jour. Soc. Tel. Eng. ii. 161.

DD. BROOKS. Lightning and Lightning Rods. Journal of the Franklin Institute, lxvi. 4.

J. PHIN. Plain Directions for the Construction of Lightning Rods. 8vo. New York.

1874. A. CALLAND. Traité des Paratonnerres—leur Utilité, leur Théorie, leur Construction. 8vo. Paris.

F. GAY-LUSSAC et CLAUDE POUILLET. Introduction sur les Paratonnerres, adoptée par l'Académie des Sciences. 8vo. Paris.

E. NOUEL. Orage du 26 Mai à Vendôme: projet de Paratonnerre simplifié. Compt. Rend. lxxix. 237.

F. MICHEL. Une Note relative à la Forme à donner aux Conducteurs des Paratonnerres. Ib. 1481.

M. MELSENS. Deuxième Note sur les Paratonnerres. Bulletin de l'Académie, Roy. de Belgique, xxxviii. 8.

M. MELSENS. Troisième Note sur les Paratonnerres. Acad. Royale de Belgique, xxxviii. 9, 10.

1875. M. MELSENS. Quatrième Note sur les Paratonnerres. Bulletin de l'Académ. Roy. de Belgique, xxxix. 6.

PRÉFET DE LA SEINE. Instruction adoptée par la Commission qui a été chargée d'étudier la meilleure disposition à donner aux Paratonnerres surmontant les Édifices municipaux et départementaux. Comp. Rend. lxxxi. 1118.

E. SAINT-EDME. Sur la Construction des Paratonnerres. Ib. 940.

J. CHEMINEAU. Une Description et un Dessin de Perfectionnement apportés aux Paratonnerres. Ib. 1203.

M. FIZEAU. Avis de la Commission des Paratonnerres sur une Disposition nouvelle proposée pour les Magasins à Poudre. Ib. lxxx. 1440.

LT. COLONEL STOTHERD, R.E. On Earth Connections of Lightning Conductors. Jour. Soc. Tel. Eng. iv. 262.

J. CLERK MAXWELL. Lightning Conductors. Ib. 429.

DR. ANTONIN DE BEAUFORT. Notice sur les Paratonnerres. 8vo. Chateauroux.

J. F. SPRAGUE. Electricity: its Theory, Sources, and Applications. 8vo. London.

DR. MANN. Lightning Conductors. Jour. Soc. of Arts. xxiii. 528.

R. F. MICHEL. On the Construction and Maintenance of Lightning Conductors. Tel. Jour. iii. 44, 63.

1876. R. F. MICHEL. Note sur la Méthode à employer pour l'Essai des Conditions de Conductibilité des Paratonnerres. Compt. Rend. lxxxii. 342.

R. F. MICHEL. Sur les Inconvénients que présente l'Emploi d'un Câble en fils de cuivre comme Conducteur de Paratonnerre. Ib. 1332.

E. SAINT-EDME. Construction of Lightning Protectors. Tel. Jour. iv. 40.

EUSTACE BUTTON. Notes on a Thunder-Storm which passed over Clevedon March 15, 1876. Jour. Soc. Tel. Eng. v. 260.

1876. W. E. AYRTON and JNO. PERRY. On Lightning Conductors. Ib. 412.
Lightning Conductors in Paris. Nature, xiii. 357.
R. J. MANN. On the Construction of Lightning Conductors. Proc. Meteorological Soc. 8vo. London.
O. BUCHNER. Die Konstruction und Anlegung der Blitzableiter, mit einem Atlas. 2nd edition. 8vo. Weimar.
Electric Conductors and Tall Chimneys. Jour. Soc. Tel. Eng. v. 531.
J. CLERK MAXWELL. On the Protection of Buildings from Lightning. Brit. Assoc. Report, 1877, 43.
R. S. NEWALL. Lightning Conductors; their use as Protectors of Buildings, and how to apply them. 8vo. London.
1877. R. WILSON. Boiler and Factory Chimneys; with a Chapter on Lightning Conductors. 8vo. London.
H. W. SPANG. A Practical Treatise on Lightning Protection. 8vo. Philadelphia.
R. S. BROUGH. On a case of Lightning, with an Evaluation of the Potential and Quantity of the Discharge in absolute Measure. Philos. Mag. vol. iv. (5 s.), 105.
DR. HELLMAN. On Thunderstorms in Central Europe. (Pogg. Ann.) 'Nature,' xv. 263.
M. JARRIANT. Sur une Nouvelle Disposition des Tiges du Paratonnerre. Compt. Rend. lxxxiv. 217.
R. S. NEWALL. Should a Lightning Conductor be insulated? Times, April 12.
M. MELSENS. De l'Application du Rhé-electromètre anx Paratonnerres des Télégraphes. Bulletins de l'Académie Royale de Belgique. xliii. No. 5.
M. MELSENS. Des Paratonnerres à pointes, à conducteurs et à raccordements terrestres multiples. 8vo. Bruxelles.
1878. M. MELSENS. Cinquième Note sur les Paratonnerres. Bulletins de l'Acad. Royale de Belgque, xlvi. No 7.
W. HOLTZ. Ueber die Theorie, die Anlage und die Prüfung der Blitzableiter. 8vo. Greifswald.
E. CARTAILHAC. Superstitious about Thunderstorms. L'Age de Pierre dans les Souvenirs et Superstitions populaires. 8vo. Paris.
R. J. MANN. Further Remarks concerning the Lightning Rod. Jour. Soc. Arts, xxvi. 328.
M. MASCARL. On Artificial Thunderstorms. Nature, xvii. 515.
R. P. BROWN. Effects of a Thunderstorm on the Colon Lighthouse. Jour. Soc. Tel. Eng. vi. 330.
PROF. C. V. ZENGER. On the Law and Origin of Thunderstorms, from the Bulletin International, Paris. Nature, xvii. 362.
RICHARD ANDERSON. On Lightning Conductors and Accidents by Lightning. The Electrician, vol. i. 215.
DR. NIPPOLDT. Dimensions of Lightning Rods. Telegraphic Journal, vi. 78.
J. B. JOULE. On a Remarkable Flash of Lightning. Nature, xviii. 260.
1879. S. A. R. On the Cause of Thunder. Nature, xx. 29.
R. S. NEWALL. On the Importance of a Sufficient Earth Contact for Lightning Conductors. Times, May 30 and June 14.
Curious Effects of Lightning. Electrician, vol. iii. 181.
AYRTON and PERRY. On the Earth Connection of Lightning Conductors. Nature, xix. 475.
G. W. CAMPHUIS. On the Effects of Lightning. Nature, xx. 96.
R. S. NEWALL. On Lightning Conductors. Nature, xx. 145.
CHARLES S. TOMES. On Lightning Conductors. Nature, xx. 145.

INDEX.

A

	PAGE
ACCIDENTS and fatalities from lightning	169–197
Admont, Styria, convent struck by lightning	67
Air-pump, the inventor of the	2
Alatri, the Cathedral of, struck by lightning	203
———————— Father Secchi's account thereof	203
Allamand (John Nicholas), his researches on electricity	4
Amber or 'Electron,' and its properties	1
America, lightning protection in	133
———— the tramping 'Lightning-rodmen' of	133
———— account of the details of the American system	134
———— utilisation of gutters and rain-pipes in	134
———— the protection of chimneys and air-shafts in	136
———— the method of constructing the earth-terminal in	136
———— the protection of mineral oil tanks	138
Antrasme, France, church twice struck by lightning at the same point	65
———————— Arago's remarks thereof	66
Arago on the observation of thunderstorms	62
———— on the efficiency of lightning-conductors	73
———— on whether lightning-conductors should be carried down inside or outside a building	159
Area of protection theory	77, 101, 126, 145
Auffangstange, the German	145
Austria, statistics of fires caused by lightning in	174

B

BADEN, statistics of deaths from lightning in	173
'Balls v. points,' the controversy of	40
Banker's iron safe struck by lightning	221
Bavaria, statistics of fires caused by lightning in	173
Becquerel (Antoine C.), his experiments on the conductivity of metals	51
Bevis (Dr.), experiments in electricity	7
Bibliography of works bearing upon lightning-conductors	231
Black Rock, Cork, St. Michael's Church struck by lightning	184
Brass wire, the use of, for lightning-conductors	105, 107
Brescia, Italy, powder-magazine destroyed by lightning	200
Brussels, the Hotel de Ville. The system of lightning-conductors at	111
Buffon (Count de), his opinion of Franklin's first pamphlet on electricity	19

	PAGE
Buffon (Count de), his promotion of experiments in electricity	20
Buttor (Eustace) account of the striking of Christ Church, Clevedon, by lightning	208

C

CARTHUSIAN monks at Paris, electrical experiments made on	6
Cavendish (Lord Charles), experiments in electricity	7
Chains, iron, the use of, for lightning-conductors	102
Chimney-shafts, the protection of, from lightning	163
Chimneys and air-shafts, the protection of, from lightning in America	136
Churches struck by lightning . 27, 38, 64, 65, 146, 147, 153, 176, 177, 181, 182, 183, 184, see also 186–196, 201, 203, 208	
Churches, the protection of, from lightning	152, 156
Coiffier first draws lightning from the atmosphere	21
Cleopatra's Needle, the protection of, from lightning	141
Clevedon, Christ Church struck by lightning	208
——— Eustace Buttor's account thereof	208
Cockburn (Sir George) and Sir William Snow Harris	89
Collinson (Peter) Correspondence with Benjamin Franklin	12, 13, 17
Compass reversed by a lightning-stroke	56
Compensator for contraction and expansion in lightning-conductors	128
Copper, the relative value of different kinds of	100
——— the necessity for its purity when used for lightning-conductors	100
——— and iron, the relative electrical conductivity of	52, 143
——— rope-conductors, the proper thickness and weight for different buildings	151
——————————— description of	162, 164
Cromer, Norfolk, church damaged by lightning	147
Cuneus, his experiments in electricity	4
Cyprus, the copper of	52

D

DALIBARD (M.), his experiments in electricity	20
Davy (Sir Humphrey), his experiments on the conductivity of metals	50
Deaths from lightning, statistics of	170–175
De la Rive (Professor) on the origin of atmospheric electricity	71
Dumdum, India, destruction of a magazine by lightning at	92

E

EARTH connection, the French methods of arranging the	131
——————————— general description of	198–217
——————————— Benjamin Franklin on	199
——————————— Rev. Dr. Hemmer on	200
——————————— Professor Landriani on	201
Electrical machines, Otto von Guericke's	2
——————— Sir Isaac Newton's	2
'Electrical tubes,' the mania for	9, 10
Electricity, the early history of	1
——— the discovery of the instantaneity of its movement	8
——— positive and negative, Benjamin Franklin on	26

	PAGE
'Electron' or amber, and its properties	1
Electro-magnetism, Hans Oersted's researches in	57
England and Wales, deaths from lightning in	170

F

FATALITIES and accidents from lightning	169–197
Fires caused by lightning in Russia	171
Folkes (Martin) experiments in electricity	7
France, the 'Instruction' of the Paris Academy on lightning-conductors	75
——— the general adoption of lightning-conductors in	77
——— the protection of powder-magazines in, from lightning	82
——— lightning protection in	125
——— neglect of lightning-conductors in	125
——— account of the details of the French system	126
——— the 'area of protection' theory in	126
——— the 'ridge-circuit' as used in	129
——— deaths from lightning in	171
Franklin (Benjamin), his early life	10, 11
————————— his first experiments in electricity	12–19
————————— correspondence thereon with Peter Collinson	12, 13, 17
————————— on the identity of lightning and electricity	16
————————— 'New Experiments and Observations in Electricity'	18
————————— his 'kite' experiment	22
————————— honours conferred on him	24
————————— his first lightning-conductor	25
————————— his experiments therewith	25
————————— on positive and negative electricity	26
————————— his lightning-conductor on West's house	30
————————— his letter to Professor Winthrop defending lightning-conductors	36
————————— his troubles in making his first lightning-conductor	101
————————— on the earth connection of lightning-conductors	100
French technical terms for lightning-conductors	102
Fuller (Thomas) on fires caused by lightning	176

G

GALVANI's experiments on animal electricity	70
Galvanometer, the invention of the	58
——————— a new form of	60
Geneva, the progress of lightning-conductors in	43
Genoa, St. Mary's Church struck by lightning	201
——————————————— Professor Landriani thereon	202
'Gentleman's Magazine' *quoted*	40
George III., his opinions on lightning-conductors	41, 42
German technical terms for lightning-conductors	102
——————— theories on the earth connection	212, 214
Germany, the progress of lightning-conductors in	43
Gilbert (Dr. William), his electrical discoveries	2
Gratz, Austria, buildings struck by lightning at	68
Gray (Stephen), his researches on electricity	3
Guericke (Otto von) his electrical machine	2

H

	PAGE
HARRIS (Sir William Snow) his efforts for the protection of ships from the effects of lightning	85
———— and Sir George Cockburn	89
———— his system for protecting ships	90
———— his 'Instructions for powder-magazines'	93
———— his system for the protection of Westminster Palace	98, 118
Hauksbee (Francis) his researches on electricity	2
Height of lightning-clouds	67
Hemmer (Rev. Dr.), his theories on the earth connection	200
Henly's system for protecting ships from lightning	90
'Heretical-rods'	44
Highbury Barn, electrical experiments made at	8
Holtz (Dr. W.), on the construction and maintenance of lightning-conductors	223
Humboldt (Alex. von) on the height of lightning-clouds	67

I

INDIA, the use of lightning-conductors in	92
Ingenhousz (Dr. Johan) and lightning-conductors	47
Inspection of lightning-conductors	218-229
'Instruction' of the Paris Academy on lightning-conductors	75
Insulators, the dangers of	147, 160, 176
Iron and Copper, the relative electrical conductivity of	52, 143
——— safe, a banker's, struck by lightning	221
Italy, the progress of lightning-conductors in	44

J

JARRIANT'S system of lightning-protection	133
Josephus' account of Solomon's Temple	63

K

KANT (Immanuel) on Benjamin Franklin	24
Kastner (Professor), his report on the partial destruction of Rosstdl Church by lightning	106
Kew, lightning-conductor erected by George III. at	41
Kinnersley (Ebenezer), his lectures on lightning-conductors	27
Kite, Benjamin Franklin's experiment with	22
Kleist (Ewald George von) and the discovery of the Leyden Jar	5

L

LANDRIANI (Professor), his theories of earth protection	201
Laughton-en-le-Morthen, church damaged by lightning	153, 176
——— R. S. Newall's comments thereon	153, 154
Lead, the use of, for lightning-conductors	104
Leicester, St. George's Church struck by lightning	177
Lenz (Professor) his experiments on the conductivity of metals	52

INDEX.

Leopold, Duke of Tuscany, and lightning-conductors	44
Le Roy (David) and the protection of the Louvre from lightning	80
———— (J. B.), his theory of protecting buildings from lightning	101
Leyden Jar, the first discovery of the	5
Lightning, superstitions in regard to	63
Lightning-clouds, the height of	67
Lightning-conductors, the discovery of	17–24
———————— early experiments with	25, 33
———————— the clergy on	26
———————— Professor Winthrop's defence of	26, 27
———————— E. Kinnersley's lectures on	27
———————— 'Poor Richard's Almanac' on	28
———————— the gradual spread of	34–48
———————— Abbé Nollet's animadversions on	35, 37
———————— Franklin's reply thereto	36, 37
———————— their general use in North America	38
———————— their first erection on St. Paul's	39
———————— their progress in Germany	43
———————— Italy	44
———————— the various metals used for	50
———————— Arago on the efficiency of	73
———————— the French 'Instruction' on	75
———————— Professor Pouillet on	78
———————— for ships	85
———————— Sir William Watson's system of, for ships	87
———————— Sir William Snow Harris's system of, for ships	90
———————— F. McTaggart's opinion of	92
———————— their use in India	92
———————— the best material for	100–110
———————— German and French technical terms for	102
———————— and weather-cocks	121
———————— Jarriant's form of	133
———————— the twofold function of	142
———————— the insulation of	147, 160, 176
———————— Newall's system of	140–168
———————— should they be carried down inside or outside the building	158
———————— Professor Clerk Maxwell's theory of	164
———————— the necessity for periodically inspecting	218
———————— Dr. W. Holtz on the construction and maintenance of	223
'Lightning-rod men,' the tramping, of America	133
Lightning and thunderstorms, character of	62
———— protection, inquiries into	73–84
Line of least resistance, the	142, 148
Lisle (M. de) on the height of lightning-clouds	67
Louis XV. and experiments in electricity	6, 10
Louvre, the protection of the, from lightning	80
———— the first public building in France fitted with lightning-conductors	80

M

McTaggart (F.), his opinion of lightning-conductors	92
Magnetisation of metals by lightning	56
Magnetism and lightning, the connection between	56

Majendie (Major), report on the destruction by lightning of the powder magazine, Victoria Colliery, Burntcliffe 147
Marly-la-Ville, Dalibard's electrical experiments at 20
Matthiessen (Professor), his researches on the conductivity of copper . . 100
Maxwell (Professor Clerk, F.R.S.), his theory of lightning protection . . 164
Melsens (Professor), his system of lightning-conductors at the Hotel de Ville, Brussels 111
Merton College Chapel, Oxford, struck by lightning 182
Metals as conductors of electricity 49–61
Metals, the different conductivity of various 50–55
Michel (R. F.), his modified terminal-rod 132
Mineral oil-tanks, the protection of, from lightning in America . . 138
Mouks, Carthusian, electrical experiments made on 6
Musschenbrock (Peter Van), his researches on electricity . . . 4, 5

N

Newall (R. S., F.R.S.), his copper-rope manufactory . . . 110, 142
———— on the church at Laughton-en-le-Morthen being struck by lightning 153, 154
Newall's system of protecting buildings from lightning . . . 140
———— copper-rope conductors 162, 164
Newbury Church, Massachusetts, struck by lightning 27
New River, electrical experiments made on the 8
Newton (Sir Isaac), his electrical machine 2
Nollet (Abbé), his criticisms on Franklin's electrical experiments . 19, 35
———— his animadversions on lightning-conductors . . 35, 37
———— Franklin's reply thereto 36, 37

O

Oersted (Hans Christian) his researches in electro-magnetism . . 55, 57
Ohm (Professor), his experiments on the conductivity of metals . . 53
Ohm's law 59
Oil, mineral, tanks, the protection of, from lightning in America . . 138
Orsini family and lightning-conductors 64
Oxford, Merton College Chapel struck by lightning 182

P

Padua, the first lightning-conductor in 48
Painting lightning-conductors 129
Paratonnerres, the Paris Academy 'Instruction' on 75
Paris Academy, the 'Instruction' of the, on lightning-conductors . . 75
Paris, death of two persons by the fall of a 'tige' from steeple of St. Gervais 146
Peltier (Jean Athanase), his researches in electricity 71
'Physico-mechanical exeperiments,' Hauksbee's 3
Pliny the Elder, on the observation of thunderstorms . . . 62
'Points v. balls,' the controversy of 40
'Poor Richard's Almanac' and lightning-conductors 28
Pope, the, on electrical experiments on monks 7
Pouillet (Professor Claude), his experiment on the conductivity of metals . 54
———— on lightning-conductors 78

	PAGE
Powder-magazines in France, the **protection of, from** lightning	82
——— ——— Sir William Snow Harris's instruction for protecting	93
Pringle (Sir John) his resignation of the Presidency of the Royal Society in 1777	41
Protestantism and lightning-conductors	43
Prussia, statistics of deaths from lightning in	170
Purfleet, building struck by lightning in 1777	41

R

RAREFIED air, the conductivity of	142, 149
Raven (Mr.), his house in Carolina, U.S., struck by lightning	159
——— ——— Arago's comments thereon	159
Réaumur (Rene Antoine de) Musschenbroek's letter to, on the Leyden Jar	5
'Return strokes' of lightning	70
Richmann (Professor G. W.), his experiments on electricity	31
——— ——— his death thereby	32
'Ridge Circuit' as used in France	129
Robespierre and lightning-conductors	36, 43
Roman Catholicism and lightning-conductors	42, 44
Rosenburg, Austria, church repeatedly struck by lightning at	64
Rosstall, Bavaria, church struck by lightning at	105
——— ——— Professor Kastner's report thereon	106
Royal Navy, vessels of the, destroyed by lightning	88
Royal Society and Benjamin Franklin	17
Russia, statistics of deaths from lightning in	171

S

ST. BRIDE'S CHURCH, London, struck by lightning in 1764	38
——— ——— Dr. William Watson's account thereof	39
——— ——— account of the damage done	183
St. Omer, the first lightning-conductor at	35
St. Paul's Cathedral, the erection of lightning-conductors upon	39–41
Saussure (Professor Horace de) erects the first lightning-conductor in Geneva	43
——— ——— the opposition thereto and his manifesto thereon	43, 44
——— ——— on the height of lightning-clouds	67
——— ——— on the origin of atmospheric electricity	70
Schleswig-Holstein, thunderstorms in	222
Secchi (Father) on the protection of churches from lightning	203
Ships destroyed by lightning, statistics of	88
Shooter's Hill, electrical experiments made at	8
Siena, the erection of lightning-conductors on the Cathedral at	45
Smoke, the conductivity of	142
Solokow and Professor Richmann's experiment in electricity	32
Solomon's Temple, its immunity from lightning-strokes	63
Staples for lightning-conductors	163
Statistics of deaths, fires, and damage caused by lightning	170
Superstitions in regard to lightning	63
Sweden, statistics of deaths from lightning in	172
Switzerland, statistics of deaths caused by lightning in	175

T

	PAGE
TERMINAL-RODS, Newall's	144
Thomson (Sir William, F.R.S.), his researches on the conductivity of copper	100
Thunderstorms and lightning, the character of	62
'Tightening-screw,' the	162
Tin, the use of, for lightning-conductors	104
Toaldo (Abbé Giuseppe) and lightning-conductors	45
'Tomlinson's Thunderstorm,' quoted	177
Torpedo fish and electric shocks	1
Trees, their liability to be struck by lightning	228
Tuscany, the erection of lightning-conductors upon powder-magazines in	48

U

UNITED STATES, lightning protection in	133
Units, the law of	58

V

VACCINATION and lightning-conductors, analogy between the progress of	46
Venice, the erection of lightning-conductors in	48
Victoria Colliery, Burntcliffe, destruction of the magazine by lightning	146
———— Major Majendie's report thereon	147
Volta and the 'return stroke'	70
Voltaire, his bon mot concerning lightning	158

W

WALL (Dr.), on electricity and lightning	3
Watson (Dr. William), experiments in electricity	7
———— the first to erect a lightning-conductor in England	38
———— on St. Bride's Church being struck by lightning	39
———— and the protection of the Royal Navy from lightning	86
Weathercocks and lightning-conductors	121
Weber (Dr.) and the law of units	59
West-End Church, Southampton, struck by lightning	181
Westminster Bridge, electrical experiments made from	7
———— Palace, the system of lightning-conductors at	98, 118
Wilson, the advocate of 'balls versus points'	40
Winckler (Dr.), his experiments in electricity	5, 6
Windsor Castle inadequately provided with lightning-conductors	175
Winthrop (Professor), his defence of lightning-conductors	26, 27
———— Franklin's letter to, defending lightning-conductors	36
Wurtemberg, statistics of deaths caused by lightning in	175

Y

YELIN (J. C. VON) his advocacy of brass wire for lightning-conductors	105

Spottiswoode & Co., Printers, New-street Square, London.